Conference Proceedings of the Society for Experimental Mechanics Series

Series Editor

Kristin B. Zimmerman
Society for Experimental Mechanics, Inc.,
Bethel, CT, USA

The Conference Proceedings of the Society for Experimental Mechanics Series presents early findings and case studies from a wide range of fundamental and applied work across the broad range of fields that comprise Experimental Mechanics. Series volumes follow the principle tracks or focus topics featured in each of the Society's two annual conferences: IMAC, A Conference and Exposition on Structural Dynamics, and the Society's Annual Conference & Exposition and will address critical areas of interest to researchers and design engineers working in all areas of Structural Dynamics, Solid Mechanics and Materials Research.

Alireza Amirkhizi • Jevan Furmanski • Christian Franck • Karen Kasza •
Aaron Forster • Jon Estrada
Editors

Challenges in Mechanics of Time-Dependent Materials & Mechanics of Biological Systems and Materials, Volume 2

Proceedings of the 2022 Annual Conference on Experimental and Applied Mechanics

Springer

Editors
Alireza Amirkhizi
UML North Campus, Dandeneau Hall 219
University of Massachusetts Lowell
Lowell, MA, USA

Jevan Furmanski
University of Dayton Research Institute
Annandale, NJ, USA

Christian Franck
UW Madison
Madison, WI, USA

Karen Kasza
Columbia University
New York, NY, USA

Aaron Forster
National Institute of Standards and Technology
Gaithersburg, MD, USA

Jon Estrada
University of Michigan
Ann Arbor, MI, USA

ISSN 2191-5644 ISSN 2191-5652 (electronic)
Conference Proceedings of the Society for Experimental Mechanics Series
ISBN 978-3-031-17456-8 ISBN 978-3-031-17457-5 (eBook)
https://doi.org/10.1007/978-3-031-17457-5

This Springer imprint is published by the registered company Springer Nature Switzerland AG
The registered company address is: Gewerbestrasse 11, 6330 Cham, Switzerland

Preface

Challenges in Mechanics of Time-Dependent Materials & Mechanics of Biological Systems and Materials represents one of six volumes of technical papers to be presented at the SEM 2022 SEM Annual Conference & Exposition on Experimental and Applied Mechanics organized by the Society for Experimental Mechanics scheduled to be held in June 13–16, 2022. The complete Proceedings also includes volumes on dynamic behavior of materials; fracture, fatigue, failure, and damage evolution; advancement in optical methods, digital image correlation, and micro-and nanomechanics; mechanics of composite, hybrid, and multifunctional materials; and thermomechanics and infrared imaging, inverse problem methodologies, and mechanics of additive and advanced manufactured materials.

Each collection presents early findings from experimental and computational investigations on an important area within experimental mechanics, the mechanics of time-dependent materials, and fracture, fatigue, failure, and damage evolution being some of these areas.

The time-dependent materials track was organized to address constitutive, time (or rate)-dependent constitutive, and fracture/failure behavior of a broad range of materials systems, including prominent research in progress in both experimental and applied mechanics. Papers concentrating on both modeling and experimental aspects of time-dependent materials are included.

The biological systems and materials segment of this volume summarizes the exchange of ideas and information among scientists and engineers involved in the research and analysis of how mechanical loads interact with the structure, properties, and function of living organisms and their tissues. The scope includes experimental, imaging, numerical, and mathematical techniques and tools spanning various length and time scales. Establishing this symposium at the Annual Meeting of the Society for Experimental Mechanics provides a venue where state-of-the-art experimental methods can be leveraged in the study of biological and bio-inspired materials, traumatic brain injury, cell mechanics, and biomechanics in general. A major goal of the symposium was for participants to collaborate in the asking of fundamental questions and the development of new techniques to address bio-inspired problems in society, human health, and the natural world. The 2022 Symposium is the 12th International Symposium on the Mechanics of Biological Systems and Materials. The organizers would like to thank all the speakers and staff at SEM for enabling a successful program.

The track organizers thank the presenters, authors, and session chairs for their participation and contribution to these tracks. The support and assistance from the SEM staff is also greatly appreciated.

Lowell, MA, USA — Alireza Amirkhizi
Annandale, NJ, USA — Jevan Furmanski
Gaithersburg, MD, USA — Aaron Forster
Madison, WI, USA — Christian Franck
New York, NY, USA — Karen Kasza
Ann Arbor, MI, USA — Jon Estrada

Contents

Chapter 1
Evaluating the Effect of the Carbon Black on the Crack Growth Behavior of Natural Rubber Sheets with Large Deformation and the Strain Distribution

H. Mitamura, K. Iizuka, and S. Yoneyama

Abstract This study evaluates the effect of the carbon black (CB) on the strain distribution in the vicinity of a crack in natural rubber (NR). In addition, the difference in crack growth behavior due to difference of the amount of CB is also evaluated. For this purpose, the crack growth behavior is observed on rubber test specimens with various of the amounts of CB. Then, the images recorded with a high-speed camera are analyzed by using digital image correlation (DIC). From this experiment, the relationship between tearing energy and crack growth rate is obtained for NR. It is revealed only the specimen filled the most CB is not broken. This result suggests that the dissipation process at the crack tip differs depending on the amount of CB.

Keywords Rubber · Crack · Digital image correlation · Large deformation · Fracture mechanics

1.1 Introduction

In recent years, there has been a growing need in the automotive industry for vehicles that are less harmful to the environment. In response, the tire industry is also seeking tires with high performance and low fuel consumption. In order to reduce the burden on the environment, it is necessary to extend the life of the rubber materials used in tire products. Southern and Thomas [1] believed that wear can be expressed by crack growth. Therefore, it is necessary to consider the crack growth in order to clarify the mechanism of wear phenomenon. The most widespread theory of fracture mechanics for rubber is that of Rivlin and Thomas [2]. The tearing energy proposed by Rivlin and Thomas avoids dealing with the complex stress field at the crack tip and only discusses the energy balance of the entire system including the crack. Thus far, the fracture mechanics of rubber has been developed by avoiding the treatment of the complex stress field at the crack tip and only treating the energy balance of the entire system. However, in order to understand the nature of crack growth, it is clearly important to measure and analyze the stress and strain at the crack tip.

Recently, the analysis of crack tips has become more and more popular, and a previous study by Rublon et al. [3] observed the elongational crystallization behavior of natural rubber crack tips from wide-angle X-ray diffractometry using high-brilliance synchrotron radiation. In this experiment, the volume of the elongation crystallization at the crack tip was plotted against the tearing energy for pure rubber and 3-level CB-filled rubbers, and a single master curve was obtained regardless of the amount of CB filling. This indicates that the crack propagation behavior is dominated by the size of the elongated crystallization region at the crack tip rather than the CB filling volume. Morishita et al. [4] also investigated the macroscopic nonlinear viscoelastic behavior of amorphous elastomers under various strain fields. There are two main types of elastomers: amorphous and strain-induced crystalline. Strain-induced crystalline elastomers, such as natural rubber, undergo crystallization at the crack tip during crack propagation [5]. Therefore, natural rubber exhibits a different crack growth behavior from that of amorphous elastomers such as SBR. Natural rubber is widely used in tires because of its high crack resistance and excellent mechanical properties, but the crack growth behavior of natural rubber depending on the amount of filler has not yet been clarified.

Therefore, this study evaluates the effect of CB on crack growth behavior and strain distribution of natural rubber by using digital image correlation (DIC) [6].

H. Mitamura (✉) · K. Iizuka · S. Yoneyama
Department of Mechanical Engineering, Aoyama Gakuin University, Sagamihara-shi, Kanagawa, Japan
e-mail: c5621172@aoyama.jp; iizuka@me.aoyama.ac.jp; yoneyama@me.aoyama.ac.jp

A. Amirkhizi et al. (eds.), *Challenges in Mechanics of Time-Dependent Materials & Mechanics of Biological Systems and Materials, Volume 2*, Conference Proceedings of the Society for Experimental Mechanics Series, https://doi.org/10.1007/978-3-031-17457-5_1

1.2 TEST Method

An initial strain is applied to a natural rubber specimen. While maintaining the initial strain, the crack is introduced with scissors, and the self-propagating crack is captured by a high-speed camera. The recorded images before and after deformation are analyzed by using DIC.

A pure shear specimen with a length of 30 mm, a width of 120 mm, and a thickness of 2 mm, as shown in Fig. 1.1, is stretched by 40 mm (133%) in a tensile testing machine. Then, a crack of about 20 mm is inserted with scissors to make the crack run by itself. A random pattern is applied to the surface of the specimen to use DIC.

A tensile testing machine (Shimadzu, SURVOPULSER) is used for tensile loading, and a digital high-speed camera (KATO KOKEN, k7 – USB) and a telephoto lens (NIKON, focal length: 200 mm) are used to photograph the crack growth behavior of the specimens. The frame rate is 200 fps, and the shutter speed is 25 μs. The obtained images are 1280 × 1024 pixels (8 bit) and 1 pixel ≈ 0.094 mm. Figure 1.2 shows a schematic diagram of the specimen divided into three regions. Region (1) is defined as $0 \leq x \leq 20$ mm, Region (2) as $20 \leq x \leq 60$ mm, and Region (3) as $60 \leq x \leq 80$ mm. The analysis region is shown in Fig. 1.3. The subset size is 15 × 15 pixels, and the strain window area is 21 × 21 pixels.

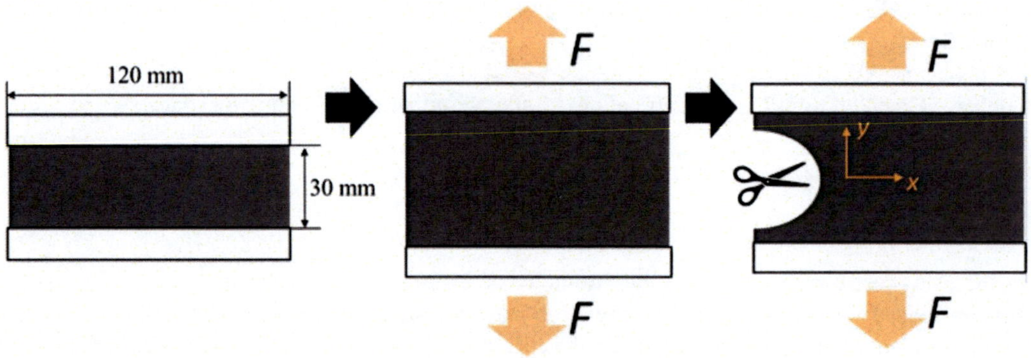

Fig. 1.1 Rubber specimen and how to experiment

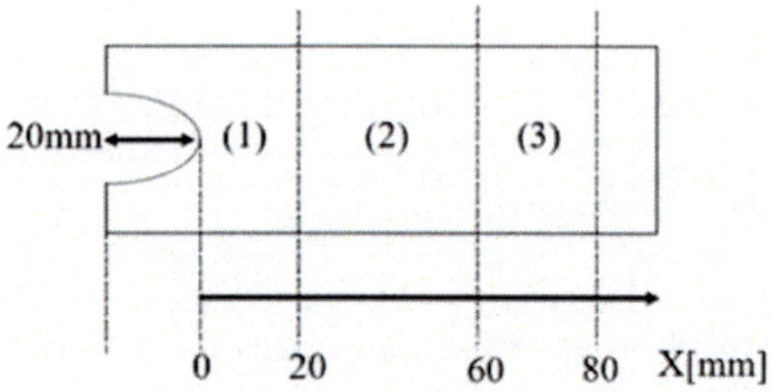

Fig. 1.2 Three regions on the test specimen

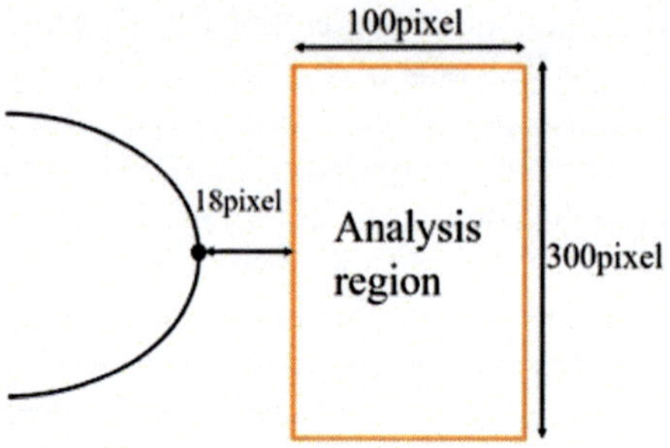

Fig. 1.3 Analysis region

1.3 Test Results and Discussion

The experimental results show all the specimens fracture except for CB50. In the case of the most CB-filled specimen CB50, the crack propagation is stopped just before rupture. It is also found that the crack propagation behavior differs depending on the amount of CB. In Fig. 1.4, it can be seen that the crack growth rate shows accelerated, constant, and decelerated behavior. The crack growth rate in the regions (1) shown in Fig. 1.2 is shown in Fig. 1.5 for CB30 and CB40. The abscissa is the position of the initial crack tip at 0 mm, and the ordinate is the crack growth rate. The acceleration of the crack growth rate after the initial crack introduction is larger in CB30 than CB40.

Figure 1.6 shows the strain distributions in the *y*-direction for CB30 and CB40 after 0.005 s using the moment of crack initiation as the reference image. In the high-speed mode, it is difficult to follow the random pattern due to the pronounced elongation near the crack, so the analysis range is shown in Fig. 1.3. From Fig. 1.6, it can be seen that the *y*-direction strain in the vicinity of a crack is higher in CB30 than in CB40.

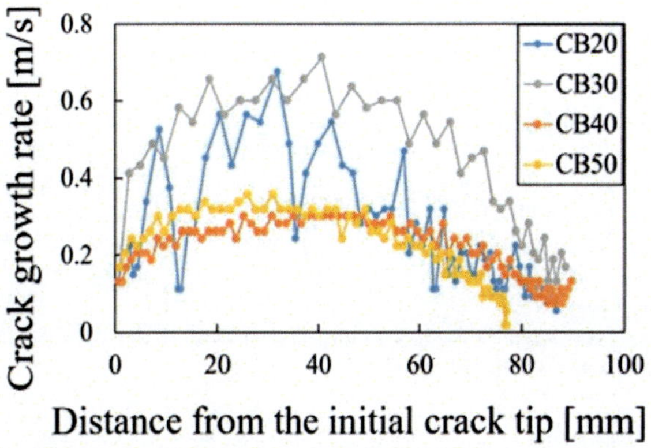

Fig. 1.4 Crack growth rate versus the distance from the initial crack tip

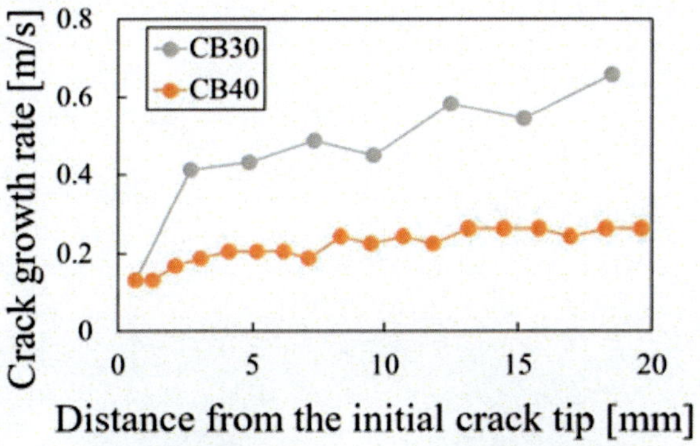

Fig. 1.5 Crack growth rate versus the distance from the initial crack tip (0~20mm)

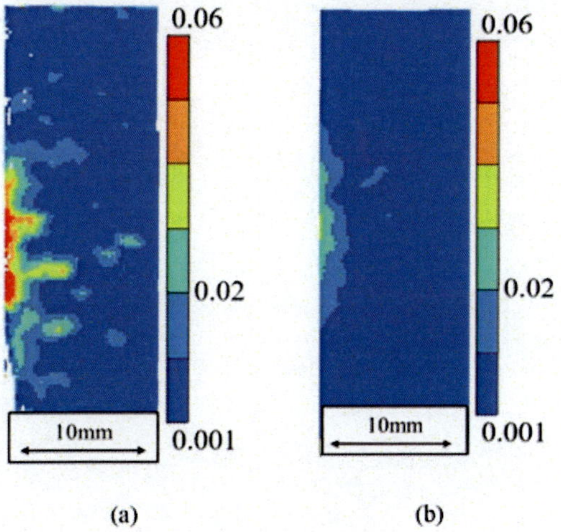

Fig. 1.6 Strain ε_y distributions (**a**) CB30, (**b**) CB40

In summary, these results suggest that the acceleration of the growth rate after the initial crack introduction is larger in CB30 than in CB40 because of the larger strain in the vicinity of a crack in CB30.

1.4 Conclusion

The crack propagation behavior is observed by applying 133% strain to rubber specimens with different amounts of CB. The experimental results show all the specimens fracture except for CB50. In the case of the most CB-filled specimen CB50, crack propagation is stopped just before rupture. It is also found that the crack propagation behavior differs depending on the amount of CB. These results suggest that the dissipation process at the crack tip differs depending on the amount of CB. For CB30 and CB40, the y-direction strain is calculated from the displacement obtained by DIC in the vicinity of a crack.

References

1. Southern, E., Thomas, A.G.: Studies of rubber abrasion. Plast. Rubber Mater. Appl. **1978**(52), 1008–1018 (1978)
2. Rivlin, R.S., Thomas, A.G.: Rupture of rubber. I. Characteristic energy for tearing. J. Polym. Sci. **310**(3), 291–318 (1953)
3. Rublon, P., Huneau, B., Verron, E., Saintier, N., Berghezan, D.: Effect of strain-induced crystallization on fatigue crack growth resistance of natural rubber. In: Constitutive Models for Rubber [VIII], pp. 349–354. CRC Press, London (2013)
4. Morishita, Y., Tsunoda, K., Urayama, K.: Crack-tip shape in the crack-growth rate transition of filled elastomers. Polymer. **108**, 230–241 (2017)
5. Huneau, B.: Strain-induced crystallization of natural rubber: a review of x-ray diffraction investigations. Rubber Chem. Technol. **84**, 425–452 (2011)
6. Yoneyama, S.: Basic principle of image correlation method and procedure for measuring in-plane displacement and strain. J. Jpn. Soc. Compos. Mater. **40**(4), 135–145 (2014)

Chapter 2
Experimental Characterization of the Rate-Dependent Tensile Failure Behavior of Photocured Elastomer

Jialiang Tao, Yuhai Xiang, Cody Schilling, Christian Franck, A. J. Boydston, and Stephan Rudykh

Abstract Photocured elastomers are widely used as 3D-printed structural materials due to their outstanding extensibility. However, their fracture behavior under different loading rates remains poorly characterized. Here, we conducted extensive experimental tests to investigate the rate-dependent mechanical behavior of a typical photocured elastomer – TangoPlus FLX 930. We applied thermal analysis to analyze its constituents and uniaxial tension tests to characterize the mechanical behavior of the elastomer from quasi-static to moderate strain rates and showed an increase in both fracture strain and stress of TangoPlus with increasing applied strain rate. To further investigate its time-dependent fracture behavior, we conducted relaxation tests under different strain and strain rates.

Keywords Rate-dependent · Fracture · Photocured elastomers

2.1 Introduction

A photocured polymer is a type of soft material that can be cured via exposure to light. Due to their flexibility, relatively low density, and high production efficiency, they are widely used in medicine, 3D printing, and photoresist production. To fully make use of its potential in various structural applications, developing failure models for this type of material is of importance. Furthermore, photocured polymers show a strong rate sensitivity and nonlinearity in almost all kinds of mechanical tests, which can be challenging when developing predictive models of its failure response.

In recent years, both theoretical and experimental efforts have been made to understand the mechanical properties of photocured polymers. For example, Wu et al. [1] investigated the evolution of the mechanical properties during photopolymerization. Zhao et al. [2] study the effect of oxygen on stress relaxation of photopolymers. Recently, Xiang et al. [3] presented a model for characterizing the light intensity-dependent viscoelasticity of the photocured polymers used in 3D printing.

However, the rate-dependent failure behavior of photocured elastomers in 3D printing applications still requires more attention. This investigation focuses on developing a micromechanical understanding of the failure behavior of the photocured elastomer, TangoPlus FLX 930, when stretched at quasi-static to moderate strain rates. First, simple uniaxial extension tests are performed in an effort to characterize the rate-dependent failure behavior at different strain rates, along with stretch and hold extension tests at designated stretch amplitudes. The results of both tests show a strong rate and and time dependency on the ultimate failure stress/strain of the material. To further evaluate the microstructure of the material, we conduct a series of experiments, including thermal analysis, recovering, and creep tests, to analyze the microstructure. Based on the self-healing behavior and viscoelasticity exhibited by this material, we develop a microstructural picture to explain the strain-rate-dependent ultimate tensile stretch, which we further validated via stress relaxation tests.

J. Tao (✉) · Y. Xiang · C. Franck (✉) · S. Rudykh
Department of Mechanical Engineering, College of Engineering, University of Wisconsin-Madison, Madison, WI, USA
e-mail: jtao22@wisc.edu; xiang33@wisc.edu; cfranck@wisc.edu; rudykh@wisc.edu

C. Schilling · A. J. Boydston
Department of Chemistry, College of Letters & Science, University of Wisconsin-Madison, Madison, WI, USA
e-mail: cschilling2@wisc.edu; aboydston@wisc.edu

A. Amirkhizi et al. (eds.), *Challenges in Mechanics of Time-Dependent Materials & Mechanics of Biological Systems and Materials, Volume 2*, Conference Proceedings of the Society for Experimental Mechanics Series, https://doi.org/10.1007/978-3-031-17457-5_2

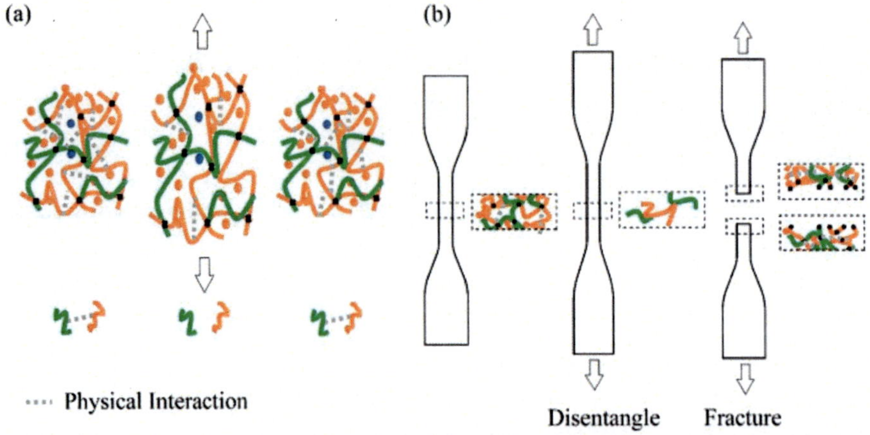

Fig. 2.1 Schematic of the proposed microstructural evolution of the TangoPlus material during the failure process. **(a)** The reversible physical interactions between chains break when the material is stretched and rebind when the specimen is unloaded. **(b)** The disentangling and breaking process of TangoPlus polymer chains during uniaxial stretching

2.2 Methods

To characterize the rate-dependent failure behavior of our TangoPlus FLX 930 elastomer, we conduct two sets of experiments: uniaxial tensile tests and stretch and hold extension tests at different engineering strain rates. The experimental setup for our uniaxial tensile and stretch and hold tests are similar to our previous arrangement for characterizing closed-cell foam material at intermediate strain rates [4]. Briefly, there dog-bone-shaped specimens with dimensions given by the ASTM D638 IV standard are produced by the Object Connex 260 3D printer. Axial strain is computed by using our previous plane fitting procedure applied to the DIC-measured, homogeneous displacement fields [5] along the axial direction. To further investigate the rate-dependent failure behavior of our photocured elastomer, a set of stretch and hold tests are also employed for a given engineering strain rate held at a designated stretch amplitude λ_c. The ultimate failure stress exhibits a strong rate and time dependency in both uniaxial tensile and stretch and hold tests (Fig. 2.1).

2.3 Microstructural Analysis

To analyze the microstructure of TangoPlus, we conducted a thermal analysis to characterize its constituents and depicted two distinct degradation temperatures, which suggest that the majority of the resin consists of at least two distinct polymer structures. Two sets of experiments are then conducted to research its recoverability. The first set is by applying a sinusoidal stretching load on the specimen with different amplitudes of applied stretch. The results show that any incurred microstructural damage can be recovered via the reversibility of physical chain interactions, i.e., entanglements or realignment among chains. The other set of experiments focuses on the self-healing ability of the polymer network. Here, immediately after an occurred tensile failure, the separated specimen pieces are rejoined and are reloaded in simple tension. Through these experiments, we observe a direct relationship of an increase in failure strain with increased self-healing recovery time. According to our experiments, we propose a microstructural picture for TangoPlus material, which consists of free chains, dangling chains, and crosslinked networks with two different chemical segments. Below the critical stress, the network can maintain structural integrity when the material is stretched at small deformation, where free chains and dangling chains provide most of the observed macroscale viscous and elastic properties. To validate the microstructure model, we performed various stress relaxation and creep tests, and our results seem to support the proposed microstructure model for TangoPlus.

2.4 Conclusion

In this paper, we systematically investigate the rate- and time-dependent failure behavior of TangoPlus via multiple experiments. We find that distinct mechanisms may dominate the failure behavior under uniaxial extension for high and low strain rates. Specifically, stretch and disentanglement of chains appear to be the cause of the failure at high and low strain rates, respectively. While the investigations are based on the common TangoPlus material, other photocured elastomers consisting of crosslinked networks and free chains with similar recoverable physical and chemical intermolecular interactions and entanglements should exhibit similar rate-dependent failure behavior as investigated here.

Acknowledgments The authors gratefully acknowledge support by the US Office of Naval Research under the PANTHER award number N000142112044 through Dr. Timothy Bentley, and the University of Wisconsin–Madison with funding from the Wisconsin Alumni Research Foundation.

References

1. Zhao, Z., Mu, X., Sowan, N., Pei, Y., Bowman, C.N., Qi, H.J., Fang, D.: Effects of oxygen on light activation in covalent adaptable network polymers. Soft Matter. **11**, 6134 (2015)
2. Wu, J., Zhao, Z., Hamel, C.M., Mu, X., Kuang, X., Guo, Z., Qi, H.J.: Evolution of material properties during free radical photopolymerization. J. Mech. Phys. Sol. **112**, 25 (2018)
3. Xiang, Y., Schilling, C., Arora, N., Boydston, A.J., Rudykh, S.: Mechanical characterization and constitutive modeling of visco-hyperelasticity of photocured polymers. Addit. Manuf. **36**, 101511 (2020)
4. Li, X., Tao, J., Landauer, A.K., Franck, C., Henann, D.L.: Large-deformation constitutive modeling of viscoelastic foams: application to a closed-cell foam material. J. Mech. Phys. Sol. **161**, 104807 (2022)
5. Landauer, A.K., Li, X., Franck, C., Henann, D.L.: A q-Factor-based digital image correlation algorithm (qDIC) for resolving finite deformations with degenerate speckle patterns. Exp. Mech. **58**, 815 (2018)

Chapter 3
Extending the Validation Range of Time-Temperature Superposition Models by Utilising the Heating Rate Dependence of the Glass Transition Temperature

Malvina Constantinou and David M. Williamson

Abstract A direct consequence of time-temperature superposition is that the observed glass transition temperature of polymers increases with strain rate. It also increases with heating rate, as has been observed in our experiments and by other authors. Our research has established a clear relationship between heating rates in thermal experiments and strain rates in mechanical experiments, by considering thermal expansion and mechanical strain to be equivalent quantities. The time-temperature superposition principle is most commonly used in the context of mechanical measurements, where the observation timescale is related to the strain rate, or oscillation frequency of the experiment. Thermal experiments access significantly larger timescales than mechanical experiments; therefore, they can extend the timescale range over which time-temperature superposition methods are validated. We use input from thermal expansion and calorimetric measurements to validate models used to predict the glass transition temperature observed via mechanical measurements, across an extended range of timescales.

Keywords Polymers · Glass transition · Thermal methods · Time-temperature superposition · Timescales

3.1 Introduction

Time-temperature superposition is a phenomenon seen in polymers, a consequence of their linear viscoelastic nature. One can consider temperature and time dependence of polymer properties as superpositions of one on the other [1], such that we can achieve a change in properties either by changing the temperature alone or by changing the timescale of measurement alone. It is often observed in mechanical and dielectric measurements that the strain rate or frequency dependence curves of a measured quantity shift with temperature, and equivalently the temperature dependence curves shift with frequency [2]. This principle is often used to construct a master curve of desired polymer properties, say, the storage modulus, by shifting available data, relative to a reference temperature.

Amorphous polymers go through a phase transition, during which the polymer goes from a glassy behaviour to a rubber-like behaviour, and there are large changes in properties, such as the storage modulus. This is referred to as the glass transition or the alpha transition and is associated with the onset of flow-like motions of the molecules relative to each other. The observed glass transition temperature (T_g) is seen to increase with increasing frequency in a dynamic mechanical analysis (DMA) experiment [3]. This may be explained using ideas from kinetic theory: the relaxation time of molecular motion will decrease with increasing temperature, and the glass transition can happen when the timescale of the external stimulus of the experiment is of the same order as the molecular relaxation time. Hence, at higher DMA frequencies, i.e. shorter experimental timescales, a higher temperature is needed to reach glass transition. A similar effect has been seen in experiments where the only external stimulus is heat. Namely, the measured glass transition temperature increases with heating/cooling rate in such thermal experiments measuring thermal expansion or heat capacity of the material [4, 5]. Shawe [5] pointed out that the heating rate is proportional to the cyclic frequency of a mechanical experiment, but hitherto, there has been no direct comparison of thermal experiments and mechanical experiments using the time-temperature superposition principle.

Our systematic measurements on a single-source polymer allowed us to do just that in the following presented research. By expressing all experiments in terms of strain rate, be it due to thermal expansion or because of a mechanical force applied to the material, we can compare the two regimes on the same time basis and relate them via time-temperature superposition

M. Constantinou (✉) · D. M. Williamson
Department of Physics, Cavendish Laboratory, University of Cambridge, Cambridge, UK
e-mail: mc954@cam.ac.uk; dmw28@cam.ac.uk

A. Amirkhizi et al. (eds.), *Challenges in Mechanics of Time-Dependent Materials & Mechanics of Biological Systems and Materials, Volume 2*, Conference Proceedings of the Society for Experimental Mechanics Series, https://doi.org/10.1007/978-3-031-17457-5_3

models. Thermal experiments lie in a longer timescale regime than mechanical experiments in general; hence, we have extended the range of timescales over which time-temperature superposition is traditionally used, in practice by several orders of magnitude. Below, we will outline the experimental results we used to obtain the glass transition temperature under different experimental timescales. We will show how to convert timescales, so that we have every experiment on the same time basis. Finally, we will present the results against a few examples of time-temperature superposition models that predict the rate dependence of the glass transition temperature.

3.2 Methods

The experimental methods used in this work include measurements of thermal expansion, measurements of heat capacity and mechanical oscillatory experiments using dynamic mechanical analysis (DMA) in cantilever mode. All experiments were performed on polycarbonate samples, machined from a single rod, ensuring there wouldn't be any differences in terms of the batch or the thermal history of the materials. Below, we include descriptions of each experimental setup and the methods of analysis of each data set to obtain the equivalent strain rate and glass transition temperature.

3.2.1 Coefficient of Thermal Expansion Measurement

Thermal expansion was measured via an in-house designed setup connected to an Attocube FPS3010 Fabry-Perot interferometer. The laser beam is reflected off a mirror on the top surface of the sample, and the shift in the interference pattern gives the displacement of this surface as the material thermally expands. The sample sits in an invar chassis, with a reference mirror on either side, all of height 6 mm. Both the chassis and reference mirrors have negligible coefficients of thermal expansion. Three spring-loaded sensor heads on the chassis point at the top surface mirrors of the sample and reference mirrors, making use of the three independent channels of the interferometer. We use the reference mirrors to account for changes in refractive index of air with temperature, any temperature gradients along the enclosure and any bulk motions. We take the true sample elongation, therefore, to be the recorded displacement, minus the mean of the reference displacements. By checking this approach with three reference mirrors (a zero-point error measurement), we find a displacement error of a few hundreds of nanometers at most, over the whole temperature range (123 K to 423 K), which, given we are detecting a few tens of μm of expansion in the polycarbonate, results in an error in absolute measurement of less than 1% and an uncertainty in the calculated coefficient of thermal expansion that is indistinguishable from noise. The enclosure is a copper heat exchanger, which can be heated with Eurotherm controlled electrical heaters and cooled using cold nitrogen gas. We measure temperature via a type K thermocouple inside the enclosure by the sample. The experiment is kept in a nitrogen-purged environment, as we wanted to eliminate humidity and avoid ice formation on the sensors and large changes in refractive index as water changes phase. The heat exchanger is surrounded by FOAMGLAS® T4+ insulation, which works well over a large range of temperatures.

For the measurement of glass transition temperature, we take the sample between room temperature and 423 K, at heating rates of 1, 7, 20, 60 and 120 K/h. Being mindful that thermal lag can also cause the observed glass transition to shift, simply because the sample is actually at a slightly lower temperature than what we record, and the other way around if it is cooling, we take data, both on heating and cooling, at the same rate. We take the average of the two thermal expansion coefficients to cancel out the equal and opposite effect that thermal lag has when heating and cooling at a given rate. From our raw measurements of sample length (l) with temperature (T), the linear thermal expansion coefficient (β) was calculated as follows:

$$\beta = \frac{1}{l}\frac{dl}{dT} \tag{3.1}$$

The gradient was measured by carrying out piecewise regression [6] at regular temperature intervals, such that the effect of noise would be averaged out rather than get enhanced in a numerical derivative calculation. The samples were found to expand isotropically, once an initial 'annealing' cycle was performed, to allow for any locked-in stresses to be relaxed out. Hence, we could take data for one of the three sides of the sample only and take $\alpha = 3\beta$ as the volumetric thermal expansion coefficient.

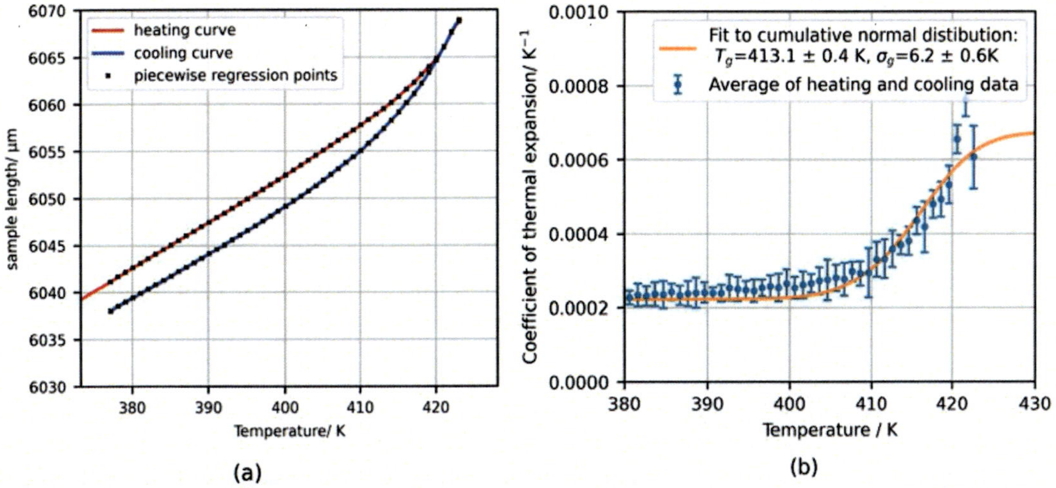

Fig. 3.1 An example of how the data is processed to obtain the glass transition temperature in a thermal expansion experiment: **(a)** raw data showing piecewise regression points, **(b)** the calculated coefficient of thermal expansion and obtaining the glass transition temperature. The experiment showed here is performed at a rate of 1 K/min

From the final coefficient of thermal expansion, we obtain the glass transition temperature, through which the value changes by a factor of 3. We have used ideas from the Group Interaction Modelling framework [7], where this change in thermal expansion coefficient results from the increase of skeletal degrees of freedom, as the chains become mobile. It is assumed that new degrees of freedom are liberated in a normal distribution with temperature, centered around T_g. We use this as justification to fit a cumulative normal distribution to α, for T_g and the distribution width. An example of the analysis is shown in Fig. 3.1.

To translate between the heating rate and equivalent thermal strain rate, we use the idea that thermal expansion itself causes a strain in the material, and the rate of this will depend on how rapidly one heats the material. Hence, from the definitions of strain rate ($\dot{\varepsilon}$) and linear thermal expansion coefficient β, we get:

$$\beta = \frac{1}{l} \cdot \frac{dl}{dT} = \frac{1}{l} \cdot \frac{dl}{dt} \cdot \frac{dt}{dT} = \frac{\dot{\varepsilon}}{Q} \tag{3.2}$$

By this simple argument, we show that the heating rate is indeed proportional to strain rate [5] and the constant of proportionality is the linear coefficient of thermal expansion. From our measurements, we find a typical value of β for glassy polycarbonate to be 7×10^{-5}, which is the value used in analysis. To express strain rate in terms of rad/s and directly compare to oscillatory mechanical experiments, we multiply the result from (3.2) by a factor of 2π.

3.2.2 Volumetric Heat Capacity: Transient Plane Source

Volumetric heat capacity was measured using a transient plane source (TPS) method [8], following ISO 22007-2:2008 [9]. A 'hot disk' sensor was sandwiched between two polycarbonate samples and acted both as a source of heating pulses and a temperature sensor. Analysis to get the volumetric heat capacity was carried out using the TPS 3500 thermal constants analyser and software. This experimental method has previously been used by our group and shown to produce reliable results by comparing measurements of well-known materials with literature values [10]. This experiment has a well-defined duration of the heat pulse, of 80s, and the temperature increase is 5K. Therefore, using Eq. (3.2), we obtain a thermal strain rate of 2.7×10^{-5} rad/s for this experiment. The glass transition temperature is defined as the middle of the step change in heat capacity, similarly to what is done above for the coefficient of thermal expansion.

3.2.3 Differential Scanning Calorimetry

A few polycarbonate samples of mass roughly 5 mg were run in a TA Instruments Q2000 DSC Machine at rates of 1, 10 and 50 K/min. The heat capacity measurement was taken both on heating and cooling to account for the effect of thermal lag. The data in these experiments included significant noise and additional effects (e.g. crystallisation on the cooling curves) so a clear step is not always obvious. The heating and cooling curves were fitted with a high-order polynomial, and the point of maximum gradient along the step change in heat capacity was taken as the glass transition temperature. The average of the heating and cooling glass transition points was taken to be the T_g for that rate, being mindful of the uncertainty involved in this measurement. The equivalent thermal strain rates for these experiments were again calculated using Eq. 3.2.

3.2.4 Dynamic Mechanical Analysis

Dynamic mechanical analysis was performed using a TA Instruments Discovery DMA 850 in a single cantilever mode on a sample of dimensions 17.5 mm length, 12.8 mm width and 1.77 mm thickness. A multifrequency temperature ramp experiment was used with a heating rate of 1 K/min and taking measurements at frequencies 0.1, 1 and 10 Hz. This results in data points roughly every 1.5 K for each frequency. The results of this method agree, within temperature uncertainty, with a reference step-wise isothermal frequency sweep, as checked using a reference sample. We therefore conclude that the effect caused by thermal lag in this experiment is no larger than 1.5 K and therefore not detectable given the temperature intervals of measurement.

The peak of loss modulus was used to identify T_g in the DMA measurements. The loss modulus is thought to be the quantity most closely related to other physical property changes due to the glass transition, such as the change in specific volume and the increase in heat capacity. If we think of T_g as the onset of chain motion, it is the loss modulus that reflects this molecular process [11]. In addition, we implement the commonly used expression [12–14] $\dot{\varepsilon} = 2\pi f \varepsilon_0$ for the strain rate, where f is the frequency of the DMA oscillation and ε_0 is the amplitude of strain as calculated from the displacement amplitude of the oscillation and the geometry of the test. This expression is sometimes used with a multiplicative factor of order unity, depending on whether the maximum or average strain rate is used. The factor will not make a significant difference for our purposes, given the logarithmic relationship of temperature and strain rate; so we take the maximum strain rate here.

Additionally, if we consider that here we have both mechanical deformation and thermal deformation, on different timescales, we will strictly have a superposition of the two rates. However, the thermal rate is several orders of magnitude smaller than the mechanical rate, so adding the two together barely makes a difference in the overall rate; mechanical rates will dominate most situations where both thermal and mechanical energy are present.

3.3 Results

Detailed results from thermal expansion experiments and the TPS volumetric heat capacity are presented elsewhere [15]. The DMA loss modulus data is plotted by frequency in Fig. 3.2.

Using the analysis outlined in the previous section, we gathered the strain rate dependence of the observed glass transition temperature. As expected from the time-temperature superposition principle, the glass transition temperature increases with increasing rates. The strain rates achieved in the thermal experiments can be multiple orders of magnitude below what can be reasonably achieved in a mechanical experiment. Hence, this is a way to extend the range of rates we validate any given time-temperature superposition model over. Discussion of model examples is included in the next section.

We summarize the results of strain rate dependence of glass transition temperatures observed in different experiments in Table 3.1. Error values for T_g reflect the temperature increments of measurements in each experiment and for DSC include the difference between the heating and cooling curves, as there is more lag and more uncertainty on where exactly T_g lies in these experiments.

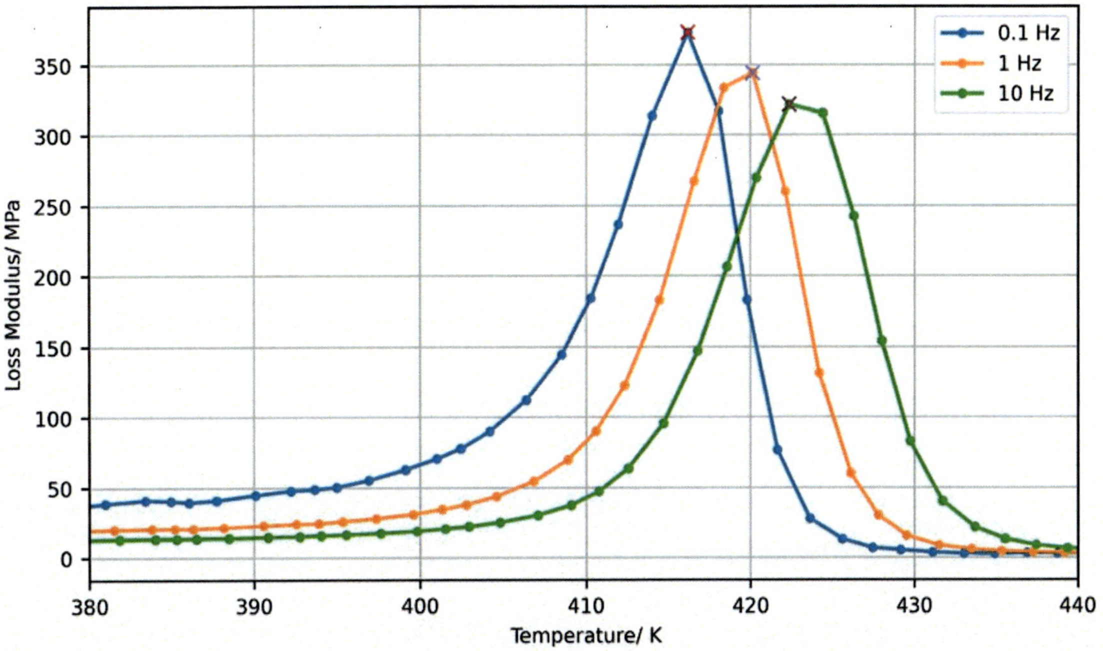

Fig. 3.2 Loss modulus measurements from the DMA test, indicating which temperature points were identified as Tg. The uncertainty reflects the temperature intervals of measurement in our experiment

Table 3.1 Summary of results from each experimental method

Thermal expansion		DSC		DMA		TPS heat capacity	
$\dot{\varepsilon}$/rad s^{-1}	T_g/K	$\dot{\varepsilon}$/rad s^{-1}	T_g/K	$\dot{\varepsilon}$/rad s^{-1}	T_g/K	$\dot{\varepsilon}$/rad s^{-1}	T_g/K
1.22×10^{-7}	410.9 ± 0.5	7.33×10^{-6}	417.2 ± 1.0	1.43×10^{-4}	416.2 ± 2.0	2.7×10^{-5}	415.0 ± 1.0
8.55×10^{-7}	411.6 ± 1.0	7.33×10^{-5}	418.3 ± 3.6	1.43×10^{-3}	420.2 ± 2.0		
2.44×10^{-6}	412.7 ± 1.0	3.66×10^{-4}	421.1 ± 6.5	1.43×10^{-2}	422.4 ± 2.0		
7.33×10^{-6}	413.1 ± 1.0						
1.46×10^{-5}	414.1 ± 2.0						

3.4 Discussion

There are many models which describe time-temperature superposition and hence can predict how the observed glass transition temperature shifts with experimental timescales. Particularly relevant to the author's research is David Porter's Group Interaction Modelling (GIM) [7] which obtains input parameters from the chemical structure of the polymer and arrives at the following equation for the T_g at a given strain rate $\dot{\varepsilon}$:

$$T_g = 0.224\theta_1 + 0.0513\frac{E_{\text{coh}}}{N} - 50 + \frac{1280 + 50\ \ln(\theta_1)}{\ln\left(\frac{2\pi f_0}{\dot{\varepsilon}}\right)} \tag{3.3}$$

This looks like the VFT [16–18] relation, but the constants are expressed in terms of the characteristic temperature (θ_1), frequency (f_0), number (N) of skeletal chain oscillations and the cohesive energy due to the interaction between neighbouring chains. The calculated parameters for polycarbonate are given in the literature [19]. We plot this prediction against our collected data in Fig. 3.3 to show that the data agrees with this model within experimental uncertainty on the wide range of timescales we have reached experimentally.

The VFT equation can be shown to be equivalent with the WLF equation. It therefore follows that if we can fit one of these relations to the data, we can fit all of them. The WLF equation [20] is expressed as follows:

$$\log_{10}\frac{\dot{\varepsilon}_0}{\dot{\varepsilon}} = \frac{-C_1\,(T - T_0)}{C_2 + (T - T_0)} \tag{3.4}$$

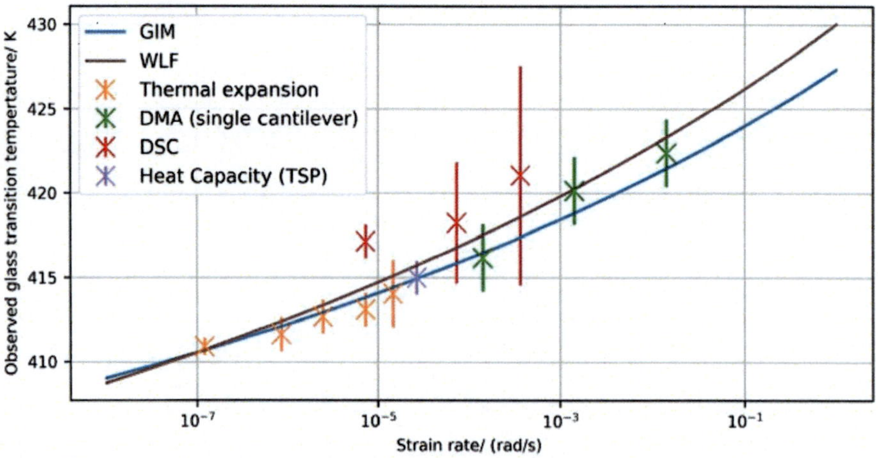

Fig. 3.3 Summary plot of all observed T_g against strain rate collected in different experiments. We show here that it is appropriate to directly compare thermal and mechanical experiments using time-temperature superposition, provided the timescale is expressed on the same basis, using the idea of a thermal strain rate caused by thermal expansion

It can be rearranged, taking the reference temperature as the glass transition temperature at strain rate $\dot{\varepsilon_0}$ and T as the observed T_g at a given strain rate. Then we get a prediction for the glass transition temperature:

$$T_g = T_{g0} - \frac{C_2 \log_{10} \frac{\dot{\varepsilon_0}}{\dot{\varepsilon}}}{\log_{10} \frac{\dot{\varepsilon_0}}{\dot{\varepsilon}} + C_1} \quad (3.5)$$

And the constants become the well-known universal constants $C_1 = 17.44$ and $C_2 = 51.6$. A fit to this form is included in Fig. 3.3, where we used the universal constants and fitted for the reference temperature and rate. This returned a reference temperature of 421 ± 3 K and a reference rate of the order 10^{-3} rad/s, which we can recognize as a typical T_g and strain rate in a 1Hz DMA experiment, which is often the choice of reference temperature in such models. We have therefore shown that the empirical WLF relation is also validated over this extended range of experimental timescales.

Note that the two models in Fig. 3.3, although they can be shown to be equivalent, are inherently different in that one (WLF) needs to be fitted to the data to get the constants and the other (GIM) is predicted entirely from the chemical structure of the polymer, without any input from the data needed. GIM clearly has more predictive power than empirical relations.

3.5 Conclusions

We have shown that it is possible to obtain a thermal strain rate, equivalent to a given heating rate in an experiment which only has heat as input energy. This is done by expressing the thermal expansion as a strain. This method allows for direct comparison of mechanical and thermal experiments on the same timescale basis. Therefore, both can be used for the validation of time-temperature superposition models, extending the validation range of such models. Inverting the argument, this also means one can predict the temperature where the glass transition will show up in a mechanical experiment while only having information from a thermal experiment. This work reinforces the origin of glass transition as the onset of molecular motions, as any measurement of the glass transition will be accompanied by deformation of the material, be it caused by thermal expansion or by a mechanical force or, indeed, by any external stimulus (e.g. one can argue this is also true in dielectric measurements).

Acknowledgements Malvina Constantinou acknowledges FOAMGLAS® for providing insulation material, Robert Cornell for performing DSC experiments and the Cavendish Laboratory Mechanical Workshop for machining samples. Acknowledgements to EPSRC and QinetiQ for funding this research.

References

1. Ferry, J.D.: Viscoelastic Properties of Polymers. Wiley, New York (1980)
2. McCrum, N.G., Read, B.E., Williams, G.: Anelastic and Dielectric Effects in Polymeric Solids. Dover Publications, New York (1967)
3. Li, R.: Time-temperature superposition method for glass transition temperature of plastic materials. Mater. Sci. Eng. A. **278**(1-2), 36–45 (2000). https://doi.org/10.1016/s0921-5093(99)00602-4
4. Greiner, R., Schwarzl, F.R.: Thermal contraction and volume relaxation of amorphous polymers. Rheol. Acta. **23**(4), 378–395 (1984). https://doi.org/10.1007/bf01329190
5. Schawe, J.E.K.: Measurement of the thermal glass transition of polystyrene in a cooling rate range of more than six decades. Thermochim. Acta. **603**, 128–134 (2015). https://doi.org/10.1016/j.tca.2014.05.025
6. Albertin, G.: Piecewise Linear Least Square Fit. MATLAB Central File Exchange. Retrieved 25 May 2020
7. Porter, D.: Group Interaction Modelling of Polymer Properties. Marcel Dekker Inc. (1995)
8. Log, T., Gustafsson, S.E.: Transient plane source (TPS) technique for measuring thermal transport properties of building materials. Fire Mater. **19**(1), 43–49 (1995). https://doi.org/10.1002/fam.810190107
9. ISO-22007-2: Plastics – Determination of Thermal Conductivity and Thermal Diffusivity -Part 2: Transient Plane Heat Source (Hot Disk) Method. International Organisation for Standardization, Geneva (2008)
10. Taylor, N.E., Williamson, D.M.: Characterising Group Interaction Modelling for Complex Composite Materials. European Office of Aerospace Research and Development, Arlington (2019)
11. Turi, E.: Thermal Characterization of Polymeric Materials, Volume I. Academic, Brooklyn/New York (2012)
12. Bartolini, L., Iannuzzi, D., Mattei, G.: Comparison of frequency and strain-rate domain mechanical characterization. Sci. Rep. **8**(1) (2018). https://doi.org/10.1038/s41598-018-31737-3
13. Kendall, M.J., Siviour, C.R.: Rate dependence of poly(vinyl chloride), the effects of plasticizer and time-temperature superposition. Proc. R. Soc. A. **470**(2167) (2014)
14. Mulliken, A.D., Boyce, M.C.: Mechanics of the rate-dependent elastic–plastic deformation of glassy polymers from low to high strain rates. Int. J. Solids Struct. **43**(5), 1331–1356 (2006). https://doi.org/10.1016/j.ijsolstr.2005.04.016
15. Constantinou, M.: *Group Interaction Modelling as a tool for Predicting Polymer Properties*. Cranfield Online Research Data (CORD) (2022). https://doi.org/10.17862/cranfield.rd.18278126.v2
16. Vogel, H.: "Das Temperaturabhaengigkeitsgesetz der Viskositaet von Fluessigkeiten" The temperature-dependent viscosity law for liquids. Physikalische Zeitschrift (in German). **22**, 645 (1921)
17. Fulcher, G.S.: Analysis of recent measurements of the viscosity of glasses. J. Am. Ceram. Soc. **8**, 339–355 (1925)
18. Tammann, G., Hesse, W.: Die Abhängigkeit der Viscosität von der Temperatur bie unterkühlten Flüssigkeiten. Z. Anorg. Allg. Chem. **156**, 245–257 (1926)
19. Porter, D., Gould, P.J.: Predictive nonlinear constitutive relations in polymers through loss history. Int. J. Solids Struct. **46**, 1981–1993 (2009)
20. Williams, M.L., Landel, R.F., Ferry, J.D.: The temperature dependence of relaxation mechanisms in amorphous polymers and other glass-forming liquids. J. Am. Chem. Soc. **77**(14), 3701–3707 (1955)

Chapter 4
Do Immobilization Methods Affect Force Spectroscopy Measurements of Single Bacteria?

Laura J. Waldman and Martha E. Grady

Abstract Increasing antibiotic resistance in bacteria is a critical issue that often leads to infections or other morbidities. Mechanical properties of the bacterial cell wall, such as thickness or elastic modulus, may contribute to the ability of a bacterial cell to resist antibiotics. Techniques like atomic force microscopy (AFM) are used to quantify bacterial cell mechanical properties and image cell structures at nanoscale resolutions. An additional benefit of AFM is the ability to probe samples submerged in liquids, meaning that live bacteria can be imaged or evaluated in environments that more accurately simulate in vivo conditions as compared to other methods like electron microscopy.

However, because AFM measurements are highly sensitive to small perturbations in the deflection of the tip of a sensor probe brought into contact with the specimen, immobilization of bacteria prior to measurement is essential for accurate measurements. Traditional chemical fixatives crosslink the molecules within the bacterial cell wall, which prevent the bacteria from locomotion. While effective for imaging, chemical crosslinkers are known to affect the measured stiffness of eukaryotic cells and also may affect the measured stiffness of the bacterial cell wall. Alternative immobilization methods include Cell-Tak™, an adhesive derived from marine mussels that does not interact with the bacterial wall and filters with known pore sizes which entrap bacteria. Previous studies have examined the effect of these immobilization methods on successful imaging of bacteria but have not addressed differences in measured modulus. This study compares the effects of immobilization methods including chemical fixatives, mechanical entrapment in filters, and Cell-Tak™ on the stiffness of the bacterial cell wall as measured by force spectroscopy.

Keywords Atomic force microscopy · Mechanobiology · Live cell immobilization

4.1 Introduction

Physical structures of bacteria are critical to the viability of cells and can affect how they move, grow, and reproduce[1–4]. Previous studies show that the elastic modulus, or stiffness, of the bacterial cell wall, may be related to the antibiotic resistance of bacteria [5, 6], which is of interest due to increasing antibiotic resistance in bacteria often leading to infections or other morbidities [7–10]. Understanding the relationship between the mechanical behavior of a cell and its antibiotic resistance may help in the development of strategies for treatment or prevention of bacterial infections.

Atomic force microscopy (AFM) is a technique used to image cell structures and quantify bacterial cell mechanical properties at high resolutions [11, 12]. One advantage to using AFM to characterize biological samples is that imaging can be performed in liquid at atmospheric pressures, simulating in vivo conditions for bacteria. Additionally, because AFM measures samples through a contact-based process, mechanical properties such as stiffness and surface roughness can also be measured. To determine the stiffness of a cell, a probe is brought into contact with the sample. The force applied to the sample and the distance the probe travels into the sample are recorded. Mathematical models, such as the Hertz fit model, are then applied to calculate properties such as elastic modulus.

However, one challenge in obtaining high-quality AFM measurements of bacterial cells is cell motility. Because AFM relies on physical contact to measure cell stiffness, the cells must be held stationary and adhered to a surface during measurement. A common method of immobilizing cells for imaging is to use a chemical crosslinker such as paraformaldehyde, which crosslinks molecules in the bacterial cell wall, "fixing" the cells. Though chemical fixatives are

L. J. Waldman (✉) · M. E. Grady
Department of Mechanical Engineering, University of Kentucky, Lexington, KY, USA
e-mail: laura.waldman@uky.edu; m.grady@uky.edu

A. Amirkhizi et al. (eds.), *Challenges in Mechanics of Time-Dependent Materials & Mechanics of Biological Systems and Materials, Volume 2*, Conference Proceedings of the Society for Experimental Mechanics Series, https://doi.org/10.1007/978-3-031-17457-5_4

effective for imaging, previous studies show that these fixatives can affect the measured Young's modulus of eukaryotic cells [13] and may affect bacterial cells as well. Previous studies which compare the effect of fixatives on bacteria focus on preserving the morphology of cell features [14, 15]. This study compares common immobilization methods to determine if the fixation method affects the measured Young's modulus of *Staphylococcus aureus* cells.

Previous studies, which use a variety of fixation methods and bacteria species, have reported modulus values between 60 kPa and 9 MPa [16–18]. However, AFM measurements are sensitive to experimental parameters such as probe tip size and shape, probe approach speed, and indention force into the sample [1, 19, 20], so a direct comparison must be made to determine the effect of a particular sample preparation method. In this study, cells will be fixed with two chemical crosslinkers, glutaraldehyde and paraformaldehyde, which are both commonly used for chemical immobilization of cells and are known to affect the viability of cells. Adhesion methods which do not kill cells include Cell-Tak™, an adhesive derived from a protein from marine mussel shells [15], and poly-L-lysine, which has a positive ionic charge and provides an attachment point for negatively charged bacterial cells [21]. However, bacteria attached with these methods can release from the surface during contact measurements. Another method used to immobilize cells for imaging is mechanical entrapment in filters with pores that correspond to cell size. *S. aureus*, which has a spherical shape, is compatible with this method, but rod-shaped bacterial species such as *Escherichia coli* cannot be immobilized in filters [15]. Future studies will compare mechanical entrapment to the chemical fixation methods as well.

4.2 Experimental

Bacteria Culture *Staphylococcus aureus* is a spherical, gram-positive bacterium. *S. aureus* stock was added to filter sterilized Todd Hewitt Yeast broth with an inoculation loop and cultured for 24 h at 37 °C.

Specimen Preparation Dish preparation was performed according to established procedure [12]. Poly-L-lysine (0.1% solution in water, Sigma-Aldrich) was added to sterile glass-bottomed 35 mm dishes to cover the bottom of the dish and left to sit for 1 h. Poly-L-lysine was aspirated from the dishes, and the dishes were rinsed three times with phosphate-buffered saline (PBS, VWR). 1 mL of *S. aureus* solution was added to each dish. After 1 h, bacterial solution was aspirated, and the dishes were rinsed three times with PBS to remove planktonic bacteria. For dishes with chemical fixatives: paraformaldehyde (4% solution in PBS, Sigma-Aldrich) or glutaraldehyde/formaldehyde (2.5% solution, Electron Microscopy Sciences) was added to the dish to cover the glass bottom and allowed to sit for an hour until bacteria was fixed. Fixative was aspirated then dishes were rinsed three times with PBS. Sample dishes were transported dry. Bacteria-coated surfaces were covered with PBS prior to imaging to simulate in vivo conditions for bacteria.

Atomic Force Microscopy Procedure Atomic force microscopy was performed with a JPK Nanowizard 4 in quantitative imaging mode combined with a Nikon confocal microscope. HQ:CSC17/Cr-AU BS cantilevers (MikroMasch) with tip radius of 8 nm and stiffness between 0.06 and 0.4 N/m were used for imaging single bacteria. Cells were imaged with a tip extend speed between 25 and 75 μm/s and a set point between 1.5 and 2 nN. Bright-field microscopy images at a magnification of 60× were taken of an area on the dish with bacterial cells prior to imaging with AFM. If the cells in the area were successfully imaged with AFM, then a second image, intended for modulus measurement, was recorded on top of one of the cells. A region from the center of this image was cropped to dimensions of 100 × 50 pixels to remove issues with measurements caused by curvature at the edges of the cell. For individual force curves, JPK data analysis software calculated Young's modulus of the cell using the Hertzian model for stress between contact of two bodies. These modulus values were averaged to provide a modulus value for an individual cell.

4.3 Results and Discussion

Representative AFM imaging and modulus results for *S. aureus* cells fixed with glutaraldehyde are shown in Fig. 4.1. Figure 4.1a shows a bright-field microscopy image of *S. aureus* cells with a 60× magnification. At this magnification, individual cells and cells that were not adhered or still moving could be visually identified. Dishes that had been treated with only poly-L-lysine had sparse cells, and cells were frequently still moving, increasing the difficulty of obtaining an accurate AFM measurement. After optical microscopy, the AFM was used to record a height map of individual bacterial cells or small

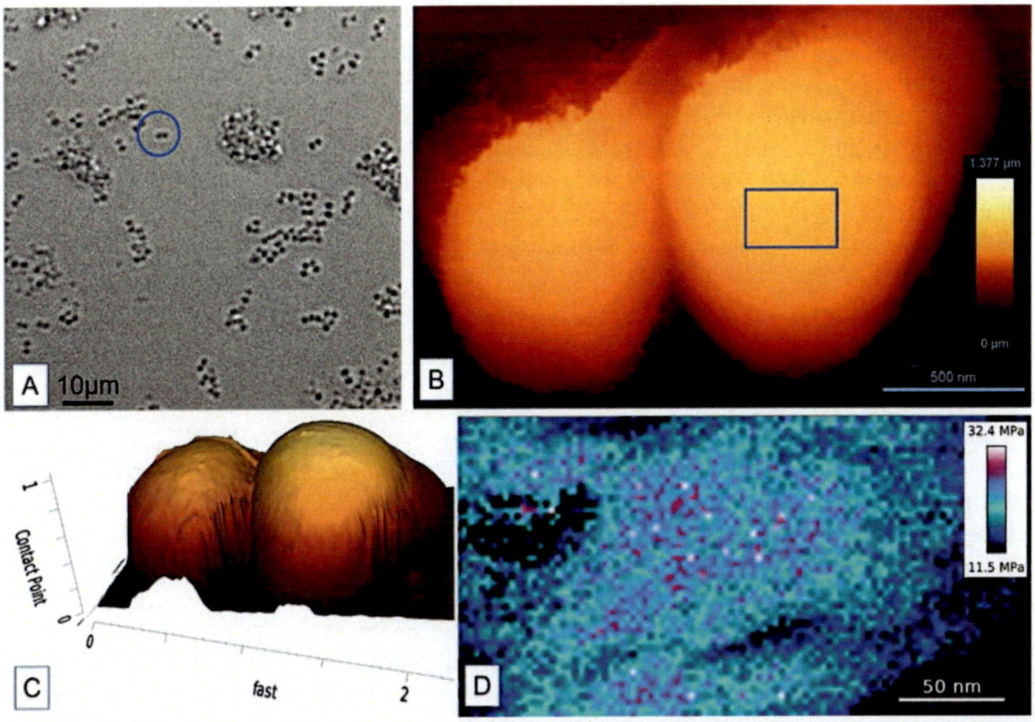

Fig. 4.1 AFM imaging and data analysis process. (**a**) 60× bright-field microscopy image of *Staphylococcus aureus* fixed with glutaraldehyde. (**b**) Atomic force microscopy map of *Staphylococcus aureus* cluster noted in (**a**). (**c**) Three-dimensional height map of the *S. aureus* cluster shown in (**b**). (**d**) Modulus map of area on top of cell highlighted in (b) that has been cropped to an area of 100 × 50 pixels

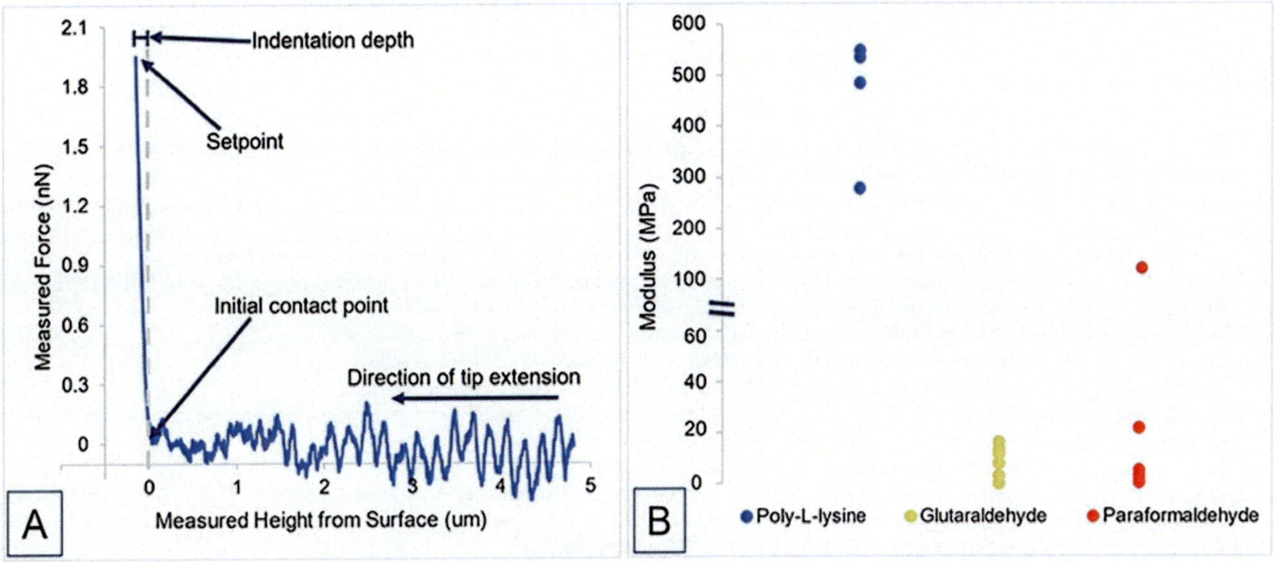

Fig. 4.2 (**a**) Representative force curve with experimental parameters labeled. In this force curve, indentation depth is 141 nm. (**b**) Experimental Young's modulus for cells with poly-L-lysine, poly-L-lysine + glutaraldehyde, and poly-L-lysine + paraformaldehyde. Note the discontinuous vertical axis

clusters of cells, as seen in Fig. 4.1b, c. No difference in cell size was noted for different fixative methods. The cells shown in these images are noted in the optical image in Fig. 4.1a.

If bacteria were successfully mapped, then a second scan was obtained from a region on top of one of the imaged cells. The approximate region chosen is shown in Fig. 4.1b. This scan was cropped to a region of 100 × 50 pixels to reduce errors caused by curvature at the edges of the cell. Each pixel represents one contact-based measurement of the cell surface, as shown in the representative force curve in Fig. 4.2a. Each force curve was fit to the Hertz model for contact stress between two bodies and used to calculate the Young's modulus at that point.

A map of modulus data for the top of one cell is shown in Fig. 4.1d. For a given map, all 5000 modulus values were averaged to determine an average value for a particular cell. Modulus values for cells adhered with poly-L-lysine and cells fixed with glutaraldehyde and paraformaldehyde are shown in Fig. 4.2b. In this preliminary study, six cells were characterized for each chemical fixative and five cells for only poly-L-lysine. Modulus values for fixed cell with both paraformaldehyde and glutaraldehyde were in a similar range to values reported in literature [16–18]. Modulus values for unfixed cells, however, were much higher than values for fixed cells and are not commonly reported in literature. Cells fixed with paraformaldehyde had the widest range in measured modulus, with values between 70 kPa and 12 MPa. Future studies will characterize more cells and explore the variation in the apparent modulus of bacteria between attachment methods. Additional immobilization techniques will also be considered.

4.4 Conclusions

Individual *Staphylococcus aureus* cells prepared with a variety of immobilizing methods were imaged with AFM. Cells without chemical fixative often detached from the surface during AFM contact measurements, increasing the difficulty of obtaining accurate measurements. Young's modulus of cells adhered with only poly-L-lysine had higher values than cells prepared with either chemical fixative. Cells prepared with paraformaldehyde had a higher range of apparent modulus values than cells prepared with glutaraldehyde. Future studies will compare additional immobilizing methods.

Acknowledgments We gratefully acknowledge NIH Center of Biomedical Research Excellence (COBRE) in Pharmaceutical Research and Innovation (CPRI, P20GM130456) and NIH NIDCR funding (R03DE029547) for completion of these experiments. AFM was performed in the Light Microscopy Core at the University of Kentucky.

Funding This material is based upon work supported by the National Science Foundation CAREER Award grant number 2045853. Any opinions, findings, and conclusions or recommendations expressed in this material are those of the authors and do not necessarily reflect the views of the National Science Foundation.

References

1. Ding, Y., et al.: Are elastic moduli of biological cells depth dependent or not? Another explanation using a contact mechanics model with surface tension. Soft Matter. **14**(36), 7534–7541 (2018)
2. Bailey, R.G., et al.: The interplay between cell wall mechanical properties and the cell cycle in staphylococcus aureus. Biophys. J. **107**(11), 2538–2545 (2014)
3. Rojas, E.R., Huang, K.C.: Regulation of microbial growth by turgor pressure. Curr. Opin. Microbiol. **42**, 62–70 (2018)
4. Wheeler, R., et al.: Bacterial cell enlargement requires control of cell wall stiffness mediated by peptidoglycan hydrolases. MBio. **6**(4), 1–10 (2015)
5. Matheli-Guinlet, M., et al.: Bacterial cell mechanics beyond Peptidoglycan. Trends Microbiol. **28**(9), 706–708 (2020)
6. Liu, L., et al.: Mechanical penetration of β-lactam-resistant Gram-negative bacteria by programmable nanowires. Sci. Adv. **6**(27), 1–12 (2020)
7. Frieri, M., Kumar, K., Boutin, A.: Antibiotic resistance. J. Infect. Publ. Health. **10**(4), 369–378 (2017)
8. Garcia-Bustos, J., Tomasz, A.: A biological price of antibiotic resistance: major changes in the peptidoglycan structure of penicillin-resistant pneumococci. Proc. Natl. Acad. Sci. U. S. A. **87**(14), 5415–5419 (1990)
9. Neu, H.C.: The crisis in antibiotic resistance. Science. **257**(5073), 1064–1073 (1992)
10. Longo, G., et al.: Antibiotic-induced modifications of the stiffness of bacterial membranes. J. Microbiol. Methods. **93**(2), 80–84 (2013)
11. Gaboriaud, F., et al.: Surface structure and nanomechanical properties of Shewanella putrefaciens bacteria at two pH values (4 and 10) determined by atomic force microscopy. J. Bacteriol. **187**(11), 3864–3868 (2005)
12. Sandin, J.N., et al.: Near simultaneous laser scanning confocal and atomic force microscopy (Conpokal) on live cells. J. Vis. Exp. **2020**(162), 1–25 (2020)
13. Riethmller, C., et al.: Vacuolar structures can be identified by AFM elasticity mapping. Ultramicroscopy. **107**(10–11), 895–901 (2007)
14. Chao, Y., Zhang, T.: Optimization of fixation methods for observation of bacterial cell morphology and surface ultrastructures by atomic force microscopy. Appl. Microbiol. Biotechnol. **92**, 381–392 (2011)
15. Meyer, R.L., et al.: Immobilisation of living bacteria for AFM imaging under physiological conditions. Ultramicroscopy. **110**(11), 1349–1357 (2010)
16. Cerf, A., et al.: Nanomechanical properties of dead or alive single-patterned bacteria. Langmuir. **25**(10), 5731–5736 (2009)
17. Chen, Y., et al.: Bacterial cell surface deformation under external loading. MBio. **3**(6) (2012)
18. Mularski, A., et al.: Atomic force microscopy reveals the mechanobiology of Lytic peptide action on bacteria. Langmuir. **31**(22), 6164–6171 (2015)
19. Kopycinska-Mller, M., Geiss, R.H., Hurley, D.C.: Contact mechanics and tip shape in AFM-based nanomechanical measurements. Ultramicroscopy. **106**(6), 466–474 (2006)
20. Ozkan, A.D., et al.: Atomic force microscopy for the investigation of molecular and cellular behavior. Micron. **89**, 60–76 (2016)
21. Colville, K., et al.: Effects of poly(L-lysine) substrates on attached escherichia coli bacteria. Langmuir. **26**(4), 2639–2644 (2010)

Chapter 5
Effect of Normal Stress on the Torsional Shear Response of the Porcine Descending Aorta

Luc Nguyen, Manoj Myneni, Abdelrahman A. Youssef, and Chandler C. Benjamin

Abstract Studying the mechanical properties of aortic tissue is important in determining future medical action regarding both healthy and diseased conditions. Although uniaxial and bi-axial experimental testing has been previously conducted on porcine aortas, shear testing is still being preliminarily explored. We studied the response of porcine descending thoracic aorta under cyclic torsional shear at five different compressive strains (5%, 10%, 15%, 20%, and 25%). From preliminary experimental results, we found that varying the normal compressive strain has a significant effect on the shear response. It can be concluded that porcine aorta should be treated as a viscoelastic material.

Keywords Shear · Compressive strain · Porcine aorta · Viscoelastic · Rheology

5.1 Introduction

In this study, we seek to characterize the response of porcine aortic tissue subjected to torsional shear with varying compressive strain applied in the direction normal to the surface of shearing. Wall shear stress and circumferential stress are significant contributing factors to arterial remodeling [7]. While the mechanics of the aortic wall has been previously studied, these studies have focused on the extensional behavior using uniaxial and bi-axial tests. Response of the aorta to shear loading has remained largely unexplored. The aorta will undergo shear deformation in various physiological situations [2], and study of the arterial wall under shear stress is integral to understanding the aorta's mechanics.

There have been previous experimental arterial studies conducted measuring the shear stress on the aorta under measured in vivo physiological conditions [1, 8]. Lu et al. [3] have conducted tri-axial shear testing on porcine thoracic aorta to experimentally determine the shear modulus of each layer of the aorta. Previous experiments in our lab on the peeling and uniaxial characteristics [4, 5] have demonstrated the nonlinear variation between the normal force and stretch placed on porcine aortas. Additionally, torsional shear testing in our lab [6] has shown a nonlinear response between the experienced torsional shear stress and the angle of twist. In the current study, we examined the rotational shear response of the porcine aorta using a rheometer under varying compressive strain. Porcine aorta was selected for testing as a precursor to potential human testing due to anatomical similarities and ease of access to a larger supply.

5.2 Experimental Methods

5.2.1 Specimen Harvest

We coordinated with the Texas A&M Animal Sciences department for the collection of porcine aortas. The aorta was cleaned of excess tissue and then sectioned into the ascending, the upper descending, and the lower descending thoracic aorta. Circular test samples were extracted from the section of the thoracic aorta between the third and the seventh branching arteries (lower descending aorta) using a half-inch circular punch. We obtained samples of uniform thickness; samples that were visually non-uniform were discarded. Uniform samples were further cleaned of excess tissue, placed in a vial filled

L. Nguyen (✉) · M. Myneni · A. A. Youssef · C. C. Benjamin
Department of Mechanical Engineering, Texas A&M University, College Station, TX, USA

A. Amirkhizi et al. (eds.), *Challenges in Mechanics of Time-Dependent Materials & Mechanics of Biological Systems and Materials, Volume 2*, Conference Proceedings of the Society for Experimental Mechanics Series, https://doi.org/10.1007/978-3-031-17457-5_5

with Phosphate Buffer Solution (PBS), then stored in a freezer maintained at −20°C where the samples remained for at least 2 weeks before testing.

5.2.2 Experimental Setup

The rheometer (Anton Paar MCR 302) was set up in the parallel plate configuration with a fixed bottom plate for a tissue bath and a 12.5 mm diameter top plate. 120-grit sandpaper was attached to both the top and bottom plates to prevent slipping of the test specimen. Before each experiment the sample was thawed by placing the frozen vial in room temperature water for at least 20 min. Measurements of the diameter and the thickness of the specimen were taken using Vernier calipers.

5.2.3 Experimental Protocol

The testing comprised of an axial normal compression followed by cyclic rotational shearing. To account for varying thicknesses of the samples, the top plate was moved down until contact with the specimen was confirmed. The bottom plate was then filled with PBS, and a hood was closed to maintain a temperature of 25°C.

An initial compression of either 5%, 10%, 15%, 20% or 25% compressive strain was applied to the specimen over the span of 1 min. After compression, the top plate rotated from +50% shear strain to −50% shear strain cyclically in a triangular waveform. 50 cycles were applied with a strain rate of 2%/s leading to a total of 5000 s of torsional shearing. The procedure was performed using a user-defined program created with the Anton Paar software.

5.3 Results

For each experiment, the rheometer recorded data for the normal force, shear strain, shear stress, and torque. Seven samples were tested at each initial compressive strain for a total of 35 samples. The data at each compressive strain was averaged and compared to show its effect on the shear response. The data presented in Fig. 5.1 shows the results of the last cycle where the shear and normal responses are nearly repeatable. The data for the shear stress response is plotted in Fig. 5.1a.

The same averaging was performed for normal force vs shear strain and torque vs shear strain plots, in Fig. 5.1b, c respectively. Shear stress is non-dimensionalized by dividing it with the mean normal stress taken as the average normal stress within the final cycle using the measured cross-section of each sample. The non-dimensional stress was then plotted against the shear strain. Figure 5.1d shows the non-dimensional stress plotted against shear strain after averaging the samples of the same initial compressive strain.

Over the course of 50 cycles for each experiment, the specimen began to experience stress-relaxation. Figure 5.2a demonstrates this by taking the average normal force for each cycle taken as the mean of all normal force values in one shear strain cycle. Figure 5.2a shows the plotted data after averaging all samples at the same initial compressive strain. Figure 5.2b does the same for the maximum shear stress recorded per cycle.

The plotted lines have individual standard deviation values which are not shown in Figs. 5.1 and 5.2. Figure 5.3 shows the averaged max shear stress per cycle for all 7 samples at 10% compressive strain along with the calculated standard deviation. This same process for calculating and plotting standard deviation can be done for all compressive strains on all plots.

5.4 Conclusions

A positive correlation between the initial normal compressive strain and the amplitude of the resultant shear stress was observed. The experiments show definite correlation between compression and shear strain, and evidence of stress-relaxation and energy dissipation. This shows that the porcine aorta responds as a viscoelastic material.

Although this experiment shows the general trend of stress-relaxation of the porcine thoracic aorta, additional relaxation focused experimentation with normal force as the independent control variable is needed to better understand the stress-

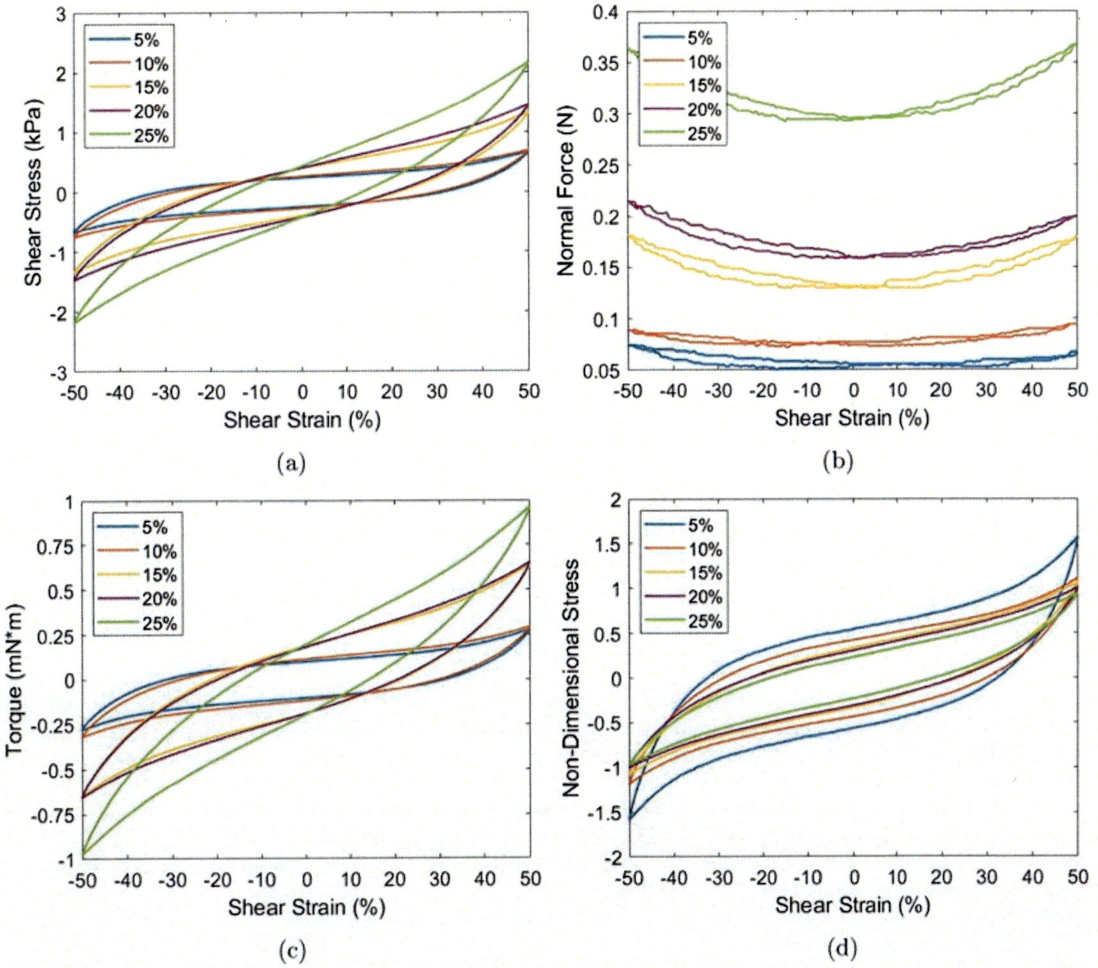

Fig. 5.1 Shear stress, normal force, torque, and non-dimensional stress as a function of shear strain. **(a)** Shear Stress vs Shear Strain. **(b)** Normal Force vs Shear Strain. **(c)** Torque vs Shear Strain. **(d)** Non-dimensional Stress vs Shear Strain

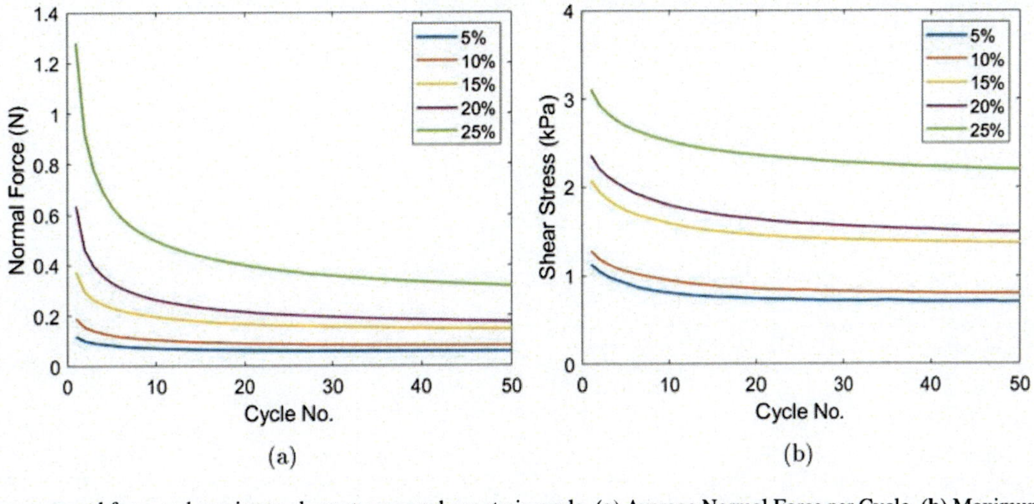

Fig. 5.2 Average normal force and maximum shear stress per shear strain cycle. **(a)** Average Normal Force per Cycle. **(b)** Maximum Shear Stress per Cycle

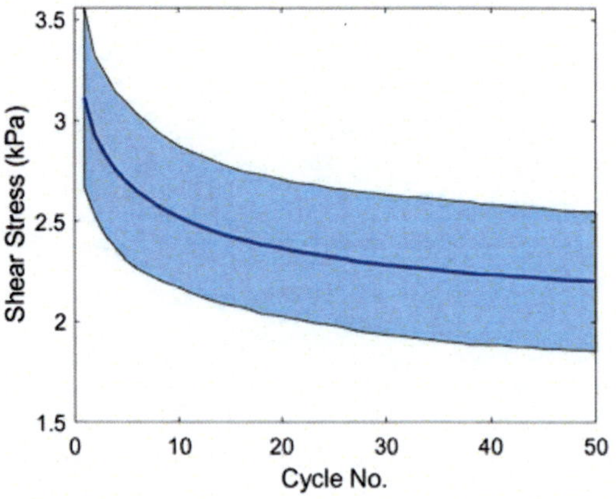

Fig. 5.3 Standard deviation of max shear stress per cycle at 10% compressive strain

relaxation response of the porcine aorta. In the future, we will focus on studying the viscoelastic shear characteristics of porcine and human aortas.

References

1. Barker, A.J., Markl, M., Bürk, J., Lorenz, R., Bock, J., Bauer, S., Schulz-Menger, J., von Knobelsdorff-Brenkenhoff, F.: Bicuspid aortic valve is associated with altered wall shear stress in the ascending aorta. Circ. Cardiovasc. Imaging **5**(4), 457–466 (2012)
2. Deng, S.X., Tomioka, J., Debes, J.C., Fung, Y.C.: New experiments on shear modulus of elasticity of arteries. Am. J. Physiol. Heart Circ. Physiol. **266**(1), H1–H10 (1994)
3. Lu, X., Yang, J., Zhao, J.B., Gregersen, H., Kassab, G.S.: Shear modulus of porcine coronary artery: contributions of media and adventitia. Am. J. Physiol. Heart Circ. Physiol. **285**(5), H1966–H1975 (2003)
4. Myneni, M., Rao, A., Jiang, M., Moreno, M.R., Rajagopal, K.R., Benjamin, C.C.: Segmental variations in the peel characteristics of the porcine thoracic aorta. Ann. Biomed. Eng. **48**(6), 1751–1767 (2020)
5. Myneni, M., Sridhar, R.L., Rajagopal, K.R., Benjamin, C.C.: Experimental investigation of the anisotropic mechanical response of the porcine thoracic aorta. Ann. Biomed. Eng. **50**(4), 452–466 (2022)
6. Rao, A., Myneni, M., Benjamin, C.C., Rajagopal, K.R.: High amplitude torsional shear of porcine thoracic aorta. In: Mechanics of Biological Systems and Materials and Micro-and Nanomechanics, vol. 4, pp. 37–40. Springer, Berlin (2020)
7. Samaee, M., Tafazzoli-Shadpour, M., Alavi, H.: Coupling of shear–circumferential stress pulses investigation through stress phase angle in FSI models of stenotic artery using experimental data. Med. Biol. Eng. Comput. **55**(8), 1147–1162 (2017)
8. Yap, C.H., Saikrishnan, N., Tamilselvan, G., Yoganathan, A.P.: Experimental measurement of dynamic fluid shear stress on the aortic surface of the aortic valve leaflet. Biomech. Model. Mechanobiol. **11**(1), 171–182 (2012)

Luc Nguyen I am a doctoral student and graduate teaching assistant at Texas A&M University in the Mechanical Engineering department with B.S. in Mechanical Engineering from Texas A&M University. I have specialized my study in biomechanics and design. I have background in experimental biomechanics and training undergraduate students in organized experimentation.

Chapter 6
A Theoretical Investigation of the Impact of Blood-Endothelium Mechanical Interactions on the Cerebral Nitric Oxide Biotransport

Corina S. Drapaca

Abstract Cerebral nitric oxide (NO), a small diffusive molecule, plays a critical role in brain's functionality. More precisely, NO acts as a neuro-glial-vascular messenger that aids various chemo-mechanical communications among brain cells, blood, and cerebral vasculature. A disequilibrium in the NO bioavailability and/or a delay in (or lack of) interactions of NO with other molecules due to, for instance, inflammation can lead to the onset and progression of cerebrovascular diseases. Mathematical models of cerebral NO biotransport can provide essential insights into mechanisms of cerebrovascular diseases that may lead to the development of better preventive, diagnostic, and therapeutic methods. In this chapter, a two-dimensional space-fractional reaction-diffusion equation is proposed to model cerebral NO biotransport. The equation uses spatial fractional order derivatives to describe NO anomalous diffusion caused by the experimentally observed entrapping of NO in circulating endothelial microparticles. Cerebrovascular diseases cause an increase in the amount of endothelial microparticles and thus may lead to an enhanced anomalous diffusion of NO. The model includes the NO production by synthesis in neurons and by shear-induced mechanotransduction in the endothelial cells, and the loss of NO due to its interactions with superoxide and hemoglobin. The blood-endothelium mechanical interactions contribute to the shear-induced production of NO. Perturbation series solutions for the coupled blood-endothelium mechanics are adapted from literature for two cases: impermeable and permeable endothelium. Thus, the viscous dissipation at the blood-endothelium interface can be calculated analytically. The model generalizes a published one-dimensional model of cerebral NO anomalous diffusion in which the blood flow effect on the vascular wall was modeled as a mere oscillatory boundary condition. Numerical simulations suggest that for NO anomalous diffusion of fractional order 0.85 and in the presence of endothelial permeability and blood flow, the NO concentration at the endothelium increases in time which agrees with studies of stroke.

Keywords Nitric oxide · Anomalous diffusion · Fractional calculus · Mechanotransduction · Blood-endothelium mechanics · Perturbation series

6.1 Introduction

Cerebral nitric oxide (NO), a small diffusive molecule, is essential to brain's functions since it acts as a neuro-glial-vascular messenger facilitating various chemo-mechanical communications among brain cells, blood, and cerebral vasculature. NO is produced by synthesis within the endothelial cells and neurons and by shear-induced mechanotransduction at the blood-endothelium interface (see [1] and references within). NO diffuses and is inactivated by some chemical processes such as its interactions with hemoglobin and superoxide. A disequilibrium in the NO bioavailability and/or a delay in (or lack of) interactions of NO with other molecules due to, for instance, inflammation or pathology can lead to the onset and progression of cerebrovascular diseases. Of relevance to this work is the fact that cerebrovascular inflammation can cause an increase in endothelial permeability [2] and in the amount of circulating endothelial microparticles (EMP) [3].

Mathematical models of cerebral NO biotransport can provide essential insights into mechanisms of cerebrovascular diseases that may lead to the development of better preventive, diagnostic, and therapeutic methods. In a recent paper [1], a one- dimensional space-fractional reaction-diffusion equation was proposed to model cerebral NO biotransport. The equation uses a spatial fractional order derivative to describe NO anomalous diffusion caused by the entrapping of NO by EMP. The model includes the NO production by synthesis in neurons and by shear-induced mechanotransduction in the endothelial

C. S. Drapaca (✉)
Department of Engineering Science and Mechanics, College of Engineering, Pennsylvania State University, University Park, PA, USA
e-mail: csd12@psu.edu

A. Amirkhizi et al. (eds.), *Challenges in Mechanics of Time-Dependent Materials & Mechanics of Biological Systems and Materials, Volume 2*, Conference Proceedings of the Society for Experimental Mechanics Series, https://doi.org/10.1007/978-3-031-17457-5_6

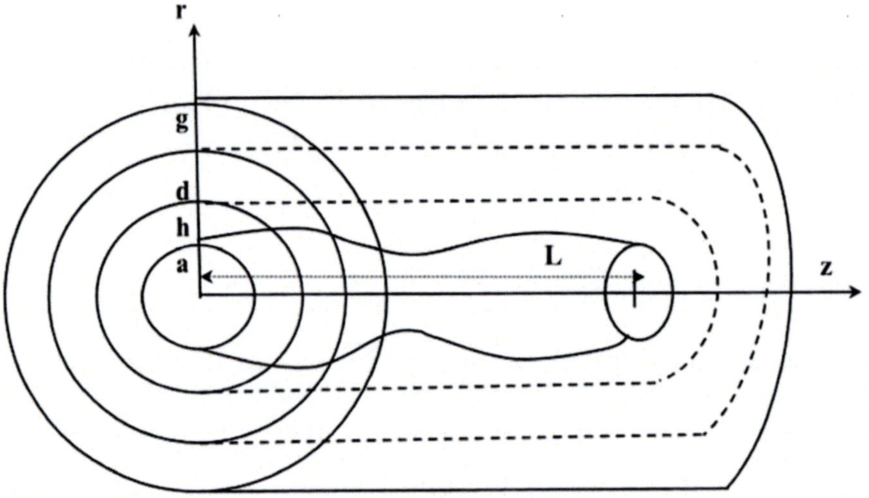

Fig. 6.1 The geometric domain is made of concentric horizontal axial symmetric circular cylinders of length L representing the lumen of radius $a = a(z, t)$, the endothelium of thickness h, a region of thickness d filled with other vascular layers and extracellular space, and a layer of thickness g occupied by neurons

cells, and the loss of NO due to its interactions with superoxide and hemoglobin. The vascular wall is assumed to be a linear viscoelastic material, and the blood flow effect on the vascular wall is modeled as a mere oscillatory boundary condition. A generalization of this model is proposed here. The model is two-dimensional, and the NO production and decay terms are like in [1]. The shear-induced production of NO is now calculated from blood-endothelium mechanical interactions. Two cases are considered: impermeable and permeable endothelium. Perturbation series solutions of a fluid-structure interaction problem that considers these two cases are given in [4, 5] and are used to calculate the viscous dissipation at the blood-endothelium interface. Numerical simulations suggest that for NO anomalous diffusion of fractional order 0.85 and in the presence of endothelial permeability and blood flow, the NO concentration at the endothelium increases in time which agrees with studies of stroke [6] showing an increased amount of endothelial NO soon after a stroke that acts as a neuroprotective agent.

6.2 Mathematical Model

Figure 6.1 shows the geometric domain of the model. It is made of concentric horizontal axial symmetric circular cylinders of length L representing the lumen of deformed radius $a = a(z, t)$, the endothelium of thickness h, a layer of thickness d filled with other vascular structures and the extracellular space, and a layer of thickness g occupied by a group of neurons. The geometry and axial symmetry assumption recommend using cylindrical coordinates $(r, z) \in [0, a_0 + h + d + g] \times [0, L]$, where a_0 is the constant undeformed radius of the lumen.

It is assumed that NO diffusion happens in the entire domain $[0, a_0 + h + d + g] \times [0, L]$, the NO production takes place in the region $([a, a + h] \cup [a + h + d, a + h + d + g]) \times [0, L]$, and the nondiffusional decay of NO happens in the region $([0, a] \cup [a + h, a + h + d + g]) \times [0, L]$. The balance law of mass for the NO concentration, $C_{NO}(r, z, t)$, has the following mathematical representation:

$$\begin{aligned}\frac{\partial C_{NO}}{\partial t} = &-\frac{1}{r}\frac{\partial}{\partial r}(rF_1) - \frac{\partial}{\partial z}F_2 + \left[\nu_1\left(1 - e^{-k_1 t}\right)e^{k_2 t}\right]H(r - (a + h + d)) - \frac{v_{\max}C_{NO}}{K_{\max}+C_{NO}}H(r - (a + h)) \\ &+ t_{rz}d_{rz}\left[H(r - a) - H(r - (a + h))\right]C_{NO} - \lambda[1 - H(r - a)]C_{NO} \\ &- \tilde{\lambda}[H(r - (a + h)) - H(r - (a + h + d))]C_{NO}\end{aligned} \tag{6.1}$$

Above, H denotes the Heaviside function.

The initial and boundary conditions used to numerically solve Eq. (6.1) are as follows:

$$C_{NO}(r,z,0)=0,\quad C_{NO}(0,z,t)=C_{NO}(a_0+h+d+g,z,t)=0,\quad C_{NO}(r,0,t)=C_{NO}(r,L,t)=0 \tag{6.2}$$

The two-dimensional diffusion flux (F_1, F_2) is given by the generalized Fick's law describing anomalous diffusion:

$$\begin{aligned} F_1(r,z,t) &= -D_{NO}\left[p_r\partial^{\varepsilon_r}{}_{L_{r+}}C_{NO}(r,z,t)-q_r\partial^{\varepsilon_r}{}_{R_{r-}}C_{NO}(r,z,t)\right], \\ F_2(r,z,t) &= -D_{NO}\left[p_z\partial^{\varepsilon_z}{}_{L_{z+}}C_{NO}(r,z,t)-q_z\partial^{\varepsilon_z}{}_{R_{z-}}C_{NO}(r,z,t)\right] \end{aligned} \tag{6.3}$$

Above, D_{NO} is a generalized diffusion coefficient, $0<\varepsilon_r,\ \varepsilon_z\le 1$ are parameters that measure the amount of long-range spatial correlations of NO particles along the radial and respectively longitudinal directions, and $p_r, q_r, p_z, q_z \ge 0$ are probabilities of moving in the forward and respectively backward diffusional radial and longitudinal directions. These probabilities satisfy the constraints $p_r+q_r=1, p_z+q_z=1$. In this chapter, it is assumed that $\varepsilon_r=\varepsilon_z=\varepsilon$, and $p_r=p_z=1$ $(q_r=q_z=0)$. The left- and right-sided Riemann-Liouville fractional derivatives of order ε with $0<\varepsilon\le 1$ are [1]:

$$\partial^{\varepsilon}_{L_{r+}}C_{NO}(r,z,t)=\frac{1}{\Gamma(1-\varepsilon)}\frac{\partial}{\partial r}\int_{L_r}^{r}\frac{C_{NO}(\tilde r,z,t)}{(r-\tilde r)^{\varepsilon}}d\tilde r,\quad \partial^{\varepsilon}_{L_{z+}}C_{NO}(r,z,t)=\frac{1}{\Gamma(1-\varepsilon)}\frac{\partial}{\partial z}\int_{L_z}^{z}\frac{C_{NO}(r,\tilde z,t)}{(z-\tilde z)^{\varepsilon}}d\tilde z \tag{6.4}$$

$$\partial^{\varepsilon}_{R_{r-}}C_{NO}(r,z,t)=\frac{1}{\Gamma(1-\varepsilon)}\frac{\partial}{\partial r}\int_{r}^{R_r}\frac{C_{NO}(\tilde r,z,t)}{(\tilde r-r)^{\varepsilon}}d\tilde r,\quad \partial^{\varepsilon}_{R_{z-}}C_{NO}(r,z,t)=\frac{1}{\Gamma(1-\varepsilon)}\frac{\partial}{\partial z}\int_{z}^{R_z}\frac{C_{NO}(r,\tilde z,t)}{(\tilde z-z)^{\varepsilon}}d\tilde z \tag{6.5}$$

where the gamma function is $\Gamma(s)=\int_0^{\infty}t^{s-1}e^{-t}dt$. It is further assumed that: $L_r=L_z=0, R_r=a_0+h+d+g, R_z=L$ Formulas (6.3) become the classic Fick's law for $\varepsilon=1$ (and $p_r=p_z=1$).

As in [1], the third term of the right-hand side of Eq. (6.1) models the time-dependent production by synthesis of the neuronal NO, while the fourth, sixth, and seventh terms represent the NO decay due to inactivation caused by NO interactions with, for instance, hemoglobin and superoxide. The fifth term models the production due to the shear-induced mechanotransduction of the endothelial NO with production rate proportional to the viscous dissipation at the blood-endothelium interface $t_{rz}d_{rz}$, where t_{rz} and d_{rz} are the nondimensional shear stress and, respectively, nondimensional shear strain rate of the blood at the blood-endothelium interface.

In conclusion, Eq. (6.1) is an extension of the one-dimensional model proposed in [1] that incorporates the anomalous diffusion of NO along the radial and longitudinal directions, the (nonconstant) deformation of the blood-endothelium interface $a=a(z,t)$, the mechanical interactions of the blood flow and endothelial deformation of an arteriole, and the endothelial permeability caused by inflammation (see later).

The fluid-structure interaction problem proposed in [4, 5] is used as a mechanical analog for the blood-endothelium mechanical interactions, and thus the perturbation series solutions given in [4, 5] are utilized to calculate the viscous dissipation and deformation $a(z,t)$ needed in Eq. (6.1). Therefore, the blood is modeled as an incompressible Newtonian viscous fluid experiencing a slow axially symmetric flow due to a known pressure gradient. The endothelium is modeled as a thin, homogeneous, isotropic, linear elastic membrane of thickness $h<<1$. The blood vessel is an arteriole that is assumed to be a long, axially symmetric cylindrical tube which is fixed at $z=0$ and stress free at $z=L$. The fluid is characterized by its mass density ρ, dynamic viscosity μ, pressure p, and the radial $v_r(r,z,t)$ and longitudinal $v_z(r,z,t)$ components of the velocity vector. The deformable membrane is characterized by its mass density ρ_m, the Young's modulus E, the Poisson's ratio v_0, and the radial $u_r(z,t)$ and longitudinal $u_z(z,t)$ components of the displacement vector. It is further assumed that:

$$\frac{a_0}{L}\ll 1,\qquad \mathrm{Re}=\frac{\rho u_* a_0}{\mu}\ll 1,\qquad \delta=\frac{\rho_m h u_*^2}{D}\ll 1,\qquad \varepsilon_m=\frac{\mu u_* L}{a_0 D}\ll 1 \tag{6.6}$$

where u_* is a characteristic longitudinal speed of the fluid and $D=Eh/\left(1-v_0^2\right)$ is the extensional rigidity of the membrane. Thus, the equation of continuity of the incompressible fluid, Navier-Stokes equations of fluid's motion, and Navier equations of membrane's motion are the governing equations of the fluid-membrane interactions [4, 5]:

$$\frac{\partial v_r}{\partial r}+\frac{v_r}{r}+\frac{\partial v_z}{\partial z}=0 \tag{6.7}$$

$$\frac{\partial v_r}{\partial t} + v_r \frac{\partial v_r}{\partial r} + v_z \frac{\partial v_r}{\partial z} = -\frac{1}{\rho}\frac{\partial p}{\partial r} + \frac{\mu}{\rho}\left(\frac{\partial^2 v_r}{\partial r^2} + \frac{1}{r}\frac{\partial v_r}{\partial r} + \frac{\partial^2 v_r}{\partial z^2} - \frac{v_r}{r^2}\right) \tag{6.8}$$

$$\frac{\partial v_z}{\partial t} + v_r \frac{\partial v_z}{\partial r} + v_z \frac{\partial v_z}{\partial z} = -\frac{1}{\rho}\frac{\partial p}{\partial z} + \frac{\mu}{\rho}\left(\frac{\partial^2 v_z}{\partial r^2} + \frac{1}{r}\frac{\partial v_z}{\partial r} + \frac{\partial^2 v_z}{\partial z^2}\right) \tag{6.9}$$

$$\rho_m h \frac{\partial^2 u_r}{\partial t^2} = -D\left(\frac{\nu_0}{a_0}\frac{\partial u_z}{\partial z} + \frac{u_r}{a_0^2}\right) + \left(p - 2\mu \frac{\partial v_r}{\partial r}\right)_{|r=a_0} \tag{6.10}$$

$$\rho_m h \frac{\partial^2 u_z}{\partial t^2} = D\left(\frac{\nu_0}{a_0}\frac{\partial u_r}{\partial z} + \frac{\partial^2 u_z}{\partial z^2}\right) - \mu\left(\frac{\partial v_z}{\partial r} + \frac{\partial v_r}{\partial z}\right)_{|r=a_0} \tag{6.11}$$

The boundary conditions are:

$$p\,(r, 0, t) = p_1(t), \qquad p\,(r, L, t) = p_2(t) \tag{6.12}$$

$$u_z\,(0, t) = 0, \qquad D\left(\frac{\partial u_z\,(L, t)}{\partial z} + \nu_0 \frac{u_r\,(L, t)}{a_0}\right) = 0 \tag{6.13}$$

In addition, if the endothelium is impermeable, then a no-slip boundary condition is imposed at $r = a(z, t) = a_0 + u_r(z, t)$ [4]:

$$v_r\,(a, z, t) = \frac{\partial u_r\,(z, t)}{\partial t},\; v_z\,(a, z, t) = \frac{\partial u_z\,(z, t)}{\partial t} \tag{6.14}$$

On the other hand, if the endothelium is permeable, then the filtration speed normal to the membrane given by Darcy's law is added to the no-slip boundary condition on $r = a(z, t)$ to get [5]:

$$v_r\,(a, z, t) = \frac{\partial u_r\,(z, t)}{\partial t} + \frac{\frac{k^*}{\mu h}\left(p\,(a, z, t) - p_e\right)}{\sqrt{1 + \left(\frac{\partial u_r}{\partial z}\right)^2}},\; v_z\,(a, z, t) = \frac{\partial u_z\,(z, t)}{\partial t} - \frac{\frac{k^*}{\mu h}\left(p\,(a, z, t) - p_e\right)}{\sqrt{1 + \left(\frac{\partial u_r}{\partial z}\right)}}\frac{\partial u_r}{\partial z} \tag{6.15}$$

where k^* is the permeability coefficient and p_e is a constant exterior pressure which, in this paper, is taken to be 0 N/m^2.

The following nondimensionalization is further introduced [4, 5]:

$$\begin{aligned} &t = \tfrac{L}{u_*}t^0, \quad z = Lz^0, \quad r = a_0 r^0, \quad a\,(z, t) = a_0 a^0\left(z^0, t^0\right), \quad u_r = \varepsilon_m a_0 u_r^0, \quad u_z = \varepsilon_m L u_z^0, \\ &v_r = \tfrac{u_* a_0}{L} v_r^0, \quad v_z = u_* v_z^0, \quad p = \tfrac{\mu u_* L}{a_0^2} p^0 \end{aligned} \tag{6.16}$$

where the superscript 0 denotes the dimensionless quantities.

By substituting formulas (6.16) into expressions (6.7)–(6.13) and using assumptions (6.6), expressions for the nondimensional components of the membrane's displacement vector and of the fluid's velocity vector can be found such that when replaced into the dimensionless forms of either boundary conditions (6.14) or boundary conditions (6.15), corresponding nonlinear differential equations for the nondimensional pressure of the fluid are obtained. The two pressure

Table 6.1 The values and physical units of model's parameters

Considerations	Parameters	Values and units [Reference]
Geometry	a_0	25×10^{-6}m [1]
	h	5×10^{-6}m [1]
	d	4×10^{-6}m [1]
	g	5×10^{-6}m [1]
	L	10^{-3}m [1]
NO diffusion, synthesis, and inactivation	D_{NO}	3.3×10^{-9}m$^{\varepsilon+1}$/s [1]
	ε	$0.65 \div 1$
	v_1	1.6×10^{-3}mol/(m^3 × s) [1]
	k_1	2s^{-1} [1]
	k_2	1.5 s^{-1} [1]
	v_{max}	2×10^{-3}mol/(m^3 × s) [1]
	K_{max}	10^{-5}mol/m^3 [1]
	λ	2.3×10^2s^{-1} [1]
	$\bar{\lambda}$	6.5325×10^3s^{-1}
Blood and endothelium mechanics	ρ_m	1 Kg/m^3 [1]
	E	6×10^3 N/m^2 [1]
	ν_0	0.5
	k_*	10^{-11} m^2
	ρ	1.06×10^3 Kg/m^3 [7]
	μ	3×10^{-3} Ns/m^2 [8]
	u_*	10^{-4} m/s
	ε_m	3×10^{-3}

equations corresponding to the two (impermeable and permeable) cases are solved using power series solutions of the form $p^0 = p^{(0)} + \varepsilon_m p^{(1)} + \frac{\varepsilon_m^2}{2!} p^{(2)} + \ldots$. Expressions for $p^{(0)}$, $p^{(1)}$ are given in [4] for the impermeable case and in [5] for the permeable case. Due to space limitations, these expressions will not be reproduced here. Thus, the following analytic expressions are found:

$$a(z,t) = a_0\left(1 + \varepsilon_m u_r^0\left(z^0, t^0\right)\right) = a_0\left(1 + \varepsilon_m\left(\frac{2-\nu_0}{2\left(1-\nu_0^2\right)} p^{(0)}\left(z^0, t^0\right) + \frac{\nu_0 p_2^0\left(t^0\right)}{2\left(1-\nu_0^2\right)}\right)\right) \tag{6.17}$$

$$t_{rz} d_{rz}\left(a^0, z^0, t^0\right)\left(= t_{rz}^0 d_{rz}^0\left(a^0, z^0, t^0\right)\right) = \left(\frac{\partial p^{(0)}}{\partial z^0}\right)^2 + 2\varepsilon_m \frac{\partial p^{(0)}}{\partial z^0}\left(\frac{\partial p^{(1)}}{\partial z^0} + \frac{2-\nu_0}{2\left(1-\nu_0^2\right)} p^{(0)} + \frac{\nu_0 p_2^0}{2\left(1-\nu_0^2\right)}\right) \tag{6.18}$$

where higher powers of ε_m are neglected thanks to assumption (6.6).

The fractional order derivatives in Eq. (6.1) are discretized using the same numerical scheme like in [1]. The space discretized problem is a system of first-order ordinary differential equations that is solved numerically using MATLAB's built-in function ode45. The nonnegative option of ode45 is chosen since negative values of C_{NO} are unphysical.

6.3 Results

The values of the model's parameters used in numerical simulations are given in Table 6.1. More details about the values of some parameters can be found in [1]. The endothelial membrane is assumed to be incompressible, and some values of parameters that could not be found in the literature are chosen such that assumptions (6.6) hold.

Since the blood vessel is assumed to be an arteriole, the pictorial of blood pressure shapes through various blood vessels given in [9] was used to propose the following expressions for the boundary conditions (6.12):

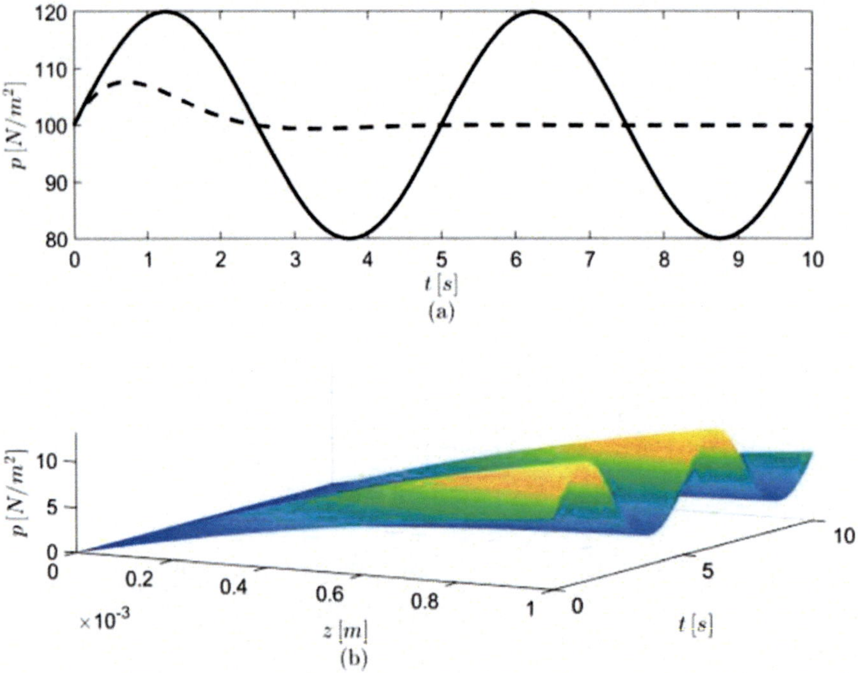

Fig. 6.2 **(a)** Graphs of the two functions given by expressions (6.19); **(b)** the spatio-temporal variations of the difference between the blood pressure calculated for a permeable endothelium [5] and the blood pressure calculated for an impermeable endothelium [4]

$$p_1(t) = 20\sin(160\pi t) + 100, \qquad p_2(t) = 20\sin(160\pi t)\exp(-t) + 100 \tag{6.19}$$

The graphs of the functions (6.19) are shown in Fig. 6.2a. The difference between the blood pressure corresponding to the permeable endothelium which was calculated in [5] and the blood pressure corresponding to the impermeable endothelium that was calculated in [4] is shown in Fig. 6.2b. The spatio-temporal distribution of this pressure difference suggests that the blood pressure increases when the endothelium becomes permeable due to inflammation.

Figures 6.3a, b, 6.4a, b, and 6.5a, b show the spatial variations of NO concentration for the impermeable endothelium and $\varepsilon = 1$ (Fig. 6.3), $\varepsilon = 0.85$ (Fig. 6.4), and $\varepsilon = 0.65$ (Fig. 6.5) at $t = 1.23$ *s* (Figs. 6.3a, 6.4a, and 6.5a) and $t = 8.75$ *s* (Figs. 6.3b, 6.4b, and 6.5b). Similarly, Figs. 6.3c, d, 6.4c, d, and 6.5c, d show the spatial variations of NO concentration for the permeable endothelium and $\varepsilon = 1$ (Fig. 6.3), $\varepsilon = 0.85$ (Fig. 6.4), and $\varepsilon = 0.65$ (Fig. 6.5) at $t = 1.23$ *s* (Figs. 6.3c, 6.4c, and 6.5c) and $t = 8.75$ *s* (Figs. 6.3d, 6.4d, and 6.5d). Differences among the patterns of NO concentration for various values of the fractional order ε as well as differences between the patterns of NO concentration for impermeable and permeable endothelium are noticeable. Lastly, Figs. 6.3e, 6.4e, and 6.5e show the upper envelops of the temporal variations of NO concentration at the endothelium and $z = L/2$ for $\varepsilon = 1$ (Fig. 6.3), $\varepsilon = 0.85$ (Fig. 6.4), and $\varepsilon = 0.65$ (Fig. 6.5). Black lines are used to represent the NO concentration for the impermeable case, and red lines are used for the permeable case. While there is very little difference between the impermeable and permeable cases for $\varepsilon = 1$ and $\varepsilon = 0.65$, when $\varepsilon = 0.85$ there is not only a clear difference between the two cases, but also it appears that the NO concentration for a permeable endothelium may increase in time. Since increased endothelial permeability of cerebral blood vessels could lead to a stroke [10], this finding agrees with published studies showing an increased amount of endothelial NO at early stages following a stroke that is needed to protect the surrounding brain cells from damage and dying [6].

6.4 Conclusion

In this chapter, a two-dimensional space-fractional reaction-diffusion equation is proposed to model cerebral NO bio-transport. Spatial fractional order derivatives are used to describe NO anomalous diffusion caused by the experimentally observed entrapping of NO in circulating endothelial microparticles. The shear-induced production of endothelial NO is assumed to have a production rate proportional to the viscous dissipation at the blood-endothelium interface. The viscous

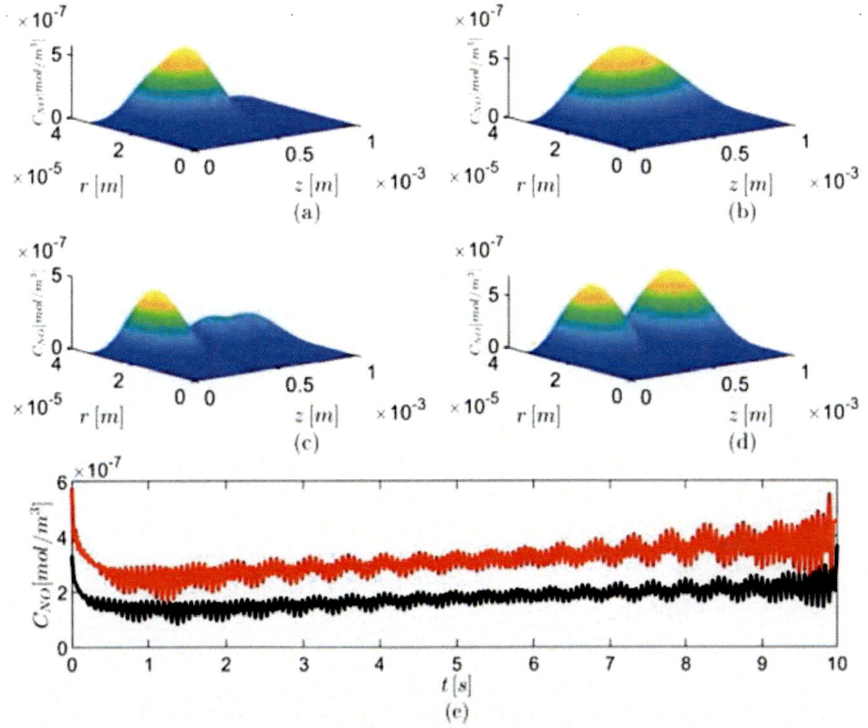

Fig. 6.3 Spatial variations of NO concentration for $\varepsilon = 1$ at $t = 1.23\ t$ (**a, c**) and $t = 8.75\ s$ (**b, d**). The impermeable case is shown on the first row (**a, b**), and the permeable case on the second row (**c, d**). (**e**) Upper envelopes of the temporal variations of the NO concentration at endothelium and $z = L/2$ for the impermeable case (black line) and permeable case (red line)

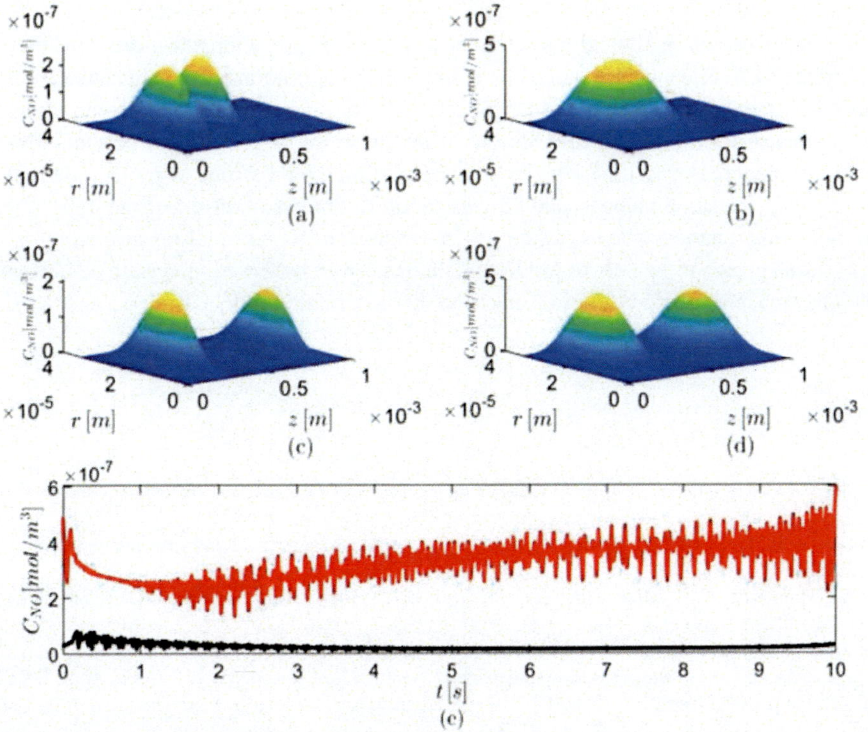

Fig. 6.4 Spatial variations of NO concentration for $\varepsilon = 0.85$ at $t = 1.23\ s$ (**a, c**) and $t = 8.75\ s$ (**b, d**). The impermeable case is shown on the first row (**a, b**), and the permeable case on the second row (**c, d**). (**e**) Upper envelopes of the temporal variations of the NO concentration at endothelium and $z = L/2$ for the impermeable case (black line) and permeable case (red line)

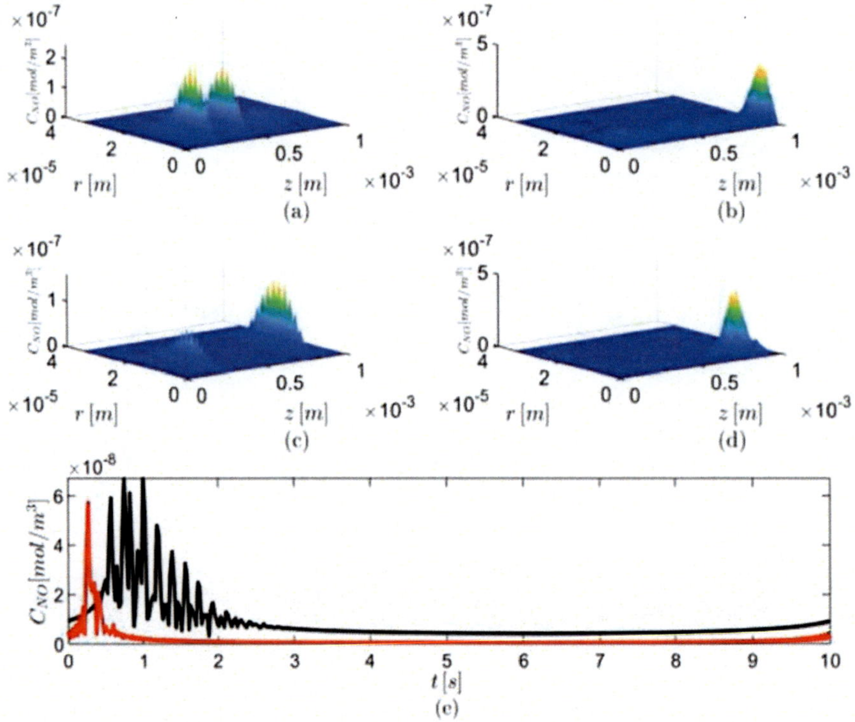

Fig. 6.5 Spatial variations of NO concentration for $\varepsilon = 0.65$ at $t = 1.23\ s$ (**a, c**) and $t = 8.75\ s$ (**b, d**). The impermeable case is shown on the first row (**a, b**), and the permeable case on the second row (**c, d**). (**e**) Upper envelopes of the temporal variations of the NO concentration at endothelium and $z = L/2$ for the impermeable case (black line) and permeable case (red line)

dissipation is calculated from known analytical solutions of a fluid-structure interaction problem that models the coupled blood- endothelium mechanics. While a healthy arteriole is impermeable, inflammation can cause an increase in endothelial permeability and thus both cases, impermeable and permeable, are considered. Numerical simulations suggest that for NO anomalous diffusion of fractional order equal to 0.85 and in the presence of endothelial permeability and blood flow, the endothelial NO concentration increases in time which can happen soon after a stroke to protect the brain cells from damage and dying. Noticeably, in [1], the same value $\varepsilon = 0.85$ was found to predict other experimentally observed links between NO amount and cerebrovascular changes due to pathology. In conclusion, NO anomalous diffusion for a characteristic jump length distribution $\varepsilon = 0.85$ appears to be able to predict NO biotransport observed in certain cerebrovascular diseases. The proposed model of NO biotransport could be further linked to more accurate models of blood-endothelium mechanics.

References

1. Tamis, A., Drapaca, C.S.: Modeling NO biotransport in brain using a space-fractional reaction-diffusion equation. Front. Physiol. **12**, 644149 (2021)
2. Claesson-Welsh, L., Dejana, E., McDonald, D.M.: Permeability of the endothelial barrier: identifying and reconciling controversies. Trends in Molecular Medicine. **27**(4), 314–331 (2021)
3. Burger, D., Schock, S., Thompson, C.S., Montezano, A.C., Hakim, A.M., Touyz, R.M.: Microparticles: biomarkers and beyond. Clinical Science. **124**(7), 423–441 (2013)
4. Camenschi, G.: A model of a viscous fluid motion through thin pipe with thin linear elastic wall. Mech. Res. Comm. **5**(5), 291–295 (1978)
5. Camenschi, G.: A mathematical model of the permeable transporting systems. Rev. Roum. Math. Pures et Appl. **XXVIII**(4), 275–282 (1983)
6. Chen, Z.-Q., Mou, R.-T., Feng, D.-X., Wang, Z., Chen, G.: The role of nitric oxide in stroke. Med. Gas. Res. **7**(3), 194–203 (2017)
7. Cutnell, J., Johnson, K.: Physics, 4th edn, p. 308. Wiley (1998)
8. Viscosity, in The Physics Hypertextbook (2022). https://physics.info/viscosity/
9. Blood Flow, Blood Pressure, and Resistance, in Anatomy and Physiology II (2022). https://courses.lumenlearning.com/suny-ap2/chapter/blood-flow-blood-pressure-and-resistance-no-content/
10. Lee, J.-M., Zhai, G., Liu, Q., Gonzales, E.R., Yin, K., Yan, P., Hsu, C.Y., Vo, K.D., Lin, W.: Vascular permeability precedes spontaneous intracerebral hemorrhage in stroke-prone spontaneously hypertensive rats. Stroke. **38**(12), 3291 (2007)

Chapter 7
The Incipient Failure of Glass Beads and Glass Bead/Epoxy Composites Under Compression as Observed by In Situ X-Ray Micro-Computed Tomography

Runyu Zhang, Christopher Paniagua, Pooyan Javadzadeh, Ning Bian, Huiluo Chen, Yao Ren, and Hongbing Lu

Abstract Granular materials are utilized in various applications, including civil engineering, energy, and defense. They exhibit very different mechanical behavior under different loading conditions. Glass beads are often used as a model granular material in simulations due to their regular geometries. In this study, we focus on understanding the incipient failure of confined glass beads assembly and a glass bead/epoxy composite under quasi-static compression using in situ X-ray micro-computed tomography (μCT). An μCT system with an integrated mechanical loading frame provides in situ volumetric images in quasi-static compression. Glass beads of various diameters (2, 3, 4, and 5 mm) are placed inside a hollow cylinder, creating an assembly consisting of glass beads with different combinations of volume fraction and particle contacts. In addition, glass bead/epoxy composite cylindrical specimens are also prepared and compressed in a separate set of experiments. The deformed configurations of such glass bead assembly and the glass bead/epoxy composite specimens at different strain levels are captured by the μCT volumetric images. The incipient failure and the damage behavior of glass beads, under the rigid confinement as well as within the epoxy binder in quasi-static compression, are visualized and assessed. This work provides experimental results for the validation of mesoscale simulations.

Keywords Granular materials · Glass beads · Particulate composite · X-ray micro-computed tomography (μCT) · Failure incipient

7.1 Introduction

Granular materials are referred to a collection of macroscopic particles. These kinds of materials are widely used in the civil and military fields; therefore, it is crucial to understand their mechanical properties upon different mechanical loading conditions. Due to the distinct and complex nature of different granular materials that may consist of particles of multiple varieties, quantities, geometries, and properties, characterizing such materials in the aggregate form is not an easy task. Many such attempts are performed experimentally and numerically, including studying the quasi-static and dynamic compressive response of sand, soils, and glass beads [1–6], and modeling and simulation work understanding the failure mechanism [7–11]. Upon loading, the formation of the force chain, which carries the load applied on the granular assembly, is highly related to the force-bearing particles and the order of the breakage of those particles [7, 11].

This work intends to supply the direct experimental measurements and observations of the incipient failure of particles in a granular assembly upon quasi-static compressive loading. The experiments are performed on a universal loading frame integrated inside an X-ray μCT system, and this in situ experimental setup captures the failure mechanism of the glass beads under confinement without a binder and also the glass bead/epoxy composite under unconfined uniaxial compression. The results provide direct observations on the interior of the granular assemblies upon failure and insights of experimental verifications for the simulation results.

R. Zhang · C. Paniagua · P. Javadzadeh · N. Bian · H. Chen · Y. Ren · H. Lu (✉)
Department of Mechanical Engineering, The University of Texas at Dallas, Richardson, TX, USA
e-mail: hongbing.lu@utdallas.edu

A. Amirkhizi et al. (eds.), *Challenges in Mechanics of Time-Dependent Materials & Mechanics of Biological Systems and Materials, Volume 2*, Conference Proceedings of the Society for Experimental Mechanics Series, https://doi.org/10.1007/978-3-031-17457-5_7

7.2 Materials and Methods

The materials used in this study were soda lime glass beads of 4 mm and 5 mm diameters (Jaygo Inc., Randolph, NJ) and a low-viscous resin system (EPIKOTE™ Resin MGS™ RIMR 135/EPIKURE™ Curing Agent MGS™ RIMH 1366, Hexion Inc.). To prepare the cylindrical glass beads embedded epoxy samples, silicone molds of ½ inch and ¾ inch inside diameter were prepared. Because the epoxy resin is relatively thick at room temperature and becomes thinner at elevated temperature, to eliminate bubbles that could potentially be generated during the sample preparation process, the premixed epoxy resin system was preheated in the oven at 70 °C for 10 minutes. Then it was slowly and carefully poured into the mold prefilled with the cleaned and dried glass beads until all glass beads were immersed. To fully eliminate the bubbles generated during the pouring process, the samples were put inside a customized ultrasonic water bath as shown in Fig. 7.1b and taken out until it was fully cured. This process proved to be effective enough to avoid any visible bubbles from occurring on the cured glass bead/epoxy composite sample shown in Fig. 7.1c, d. Since the glass bead/epoxy composite materials consist of the epoxy resin material as the matrix and the glass beads as the inclusion, it is worthwhile to explore the compressive response of each material individually. Firstly, single glass beads of the diameter of 2, 3, 4, and 5 mm were compressed using the Instron universal testing machine. After 20 compression experiments on the single glass bead of each diameter to collect statistical results, Hertzian contact stress analysis was performed. Then, the compressive response of the epoxy resin was obtained from the uniaxial compression experiments on the cylindrical neat epoxy resin samples.

To study the compressive behavior and incipient failure of the glass beads assembly under confinement, the in situ compression experiments were conducted on a Nikon 225 kV X-ray μCT system with an integrated mechanical loading frame (Psylotech, Evanston, IL) as shown in Fig. 7.2. This in situ testing system allows for μCT scans performed, while the samples are under mechanical loading (maximum 45 kN and 300 N·m for load and torque). The glass beads were washed and subsequently dried in an oven and then put inside a sapphire tube of ¼ inch inside diameter capped by cemented tungsten carbide cylindrical rods. The two rods acted as the compression platens and applied compressive load on the glass beads assembly confined by the sapphire tube. Then each μCT scan generated raw projections that had a size of 2000 x 2000 pixels

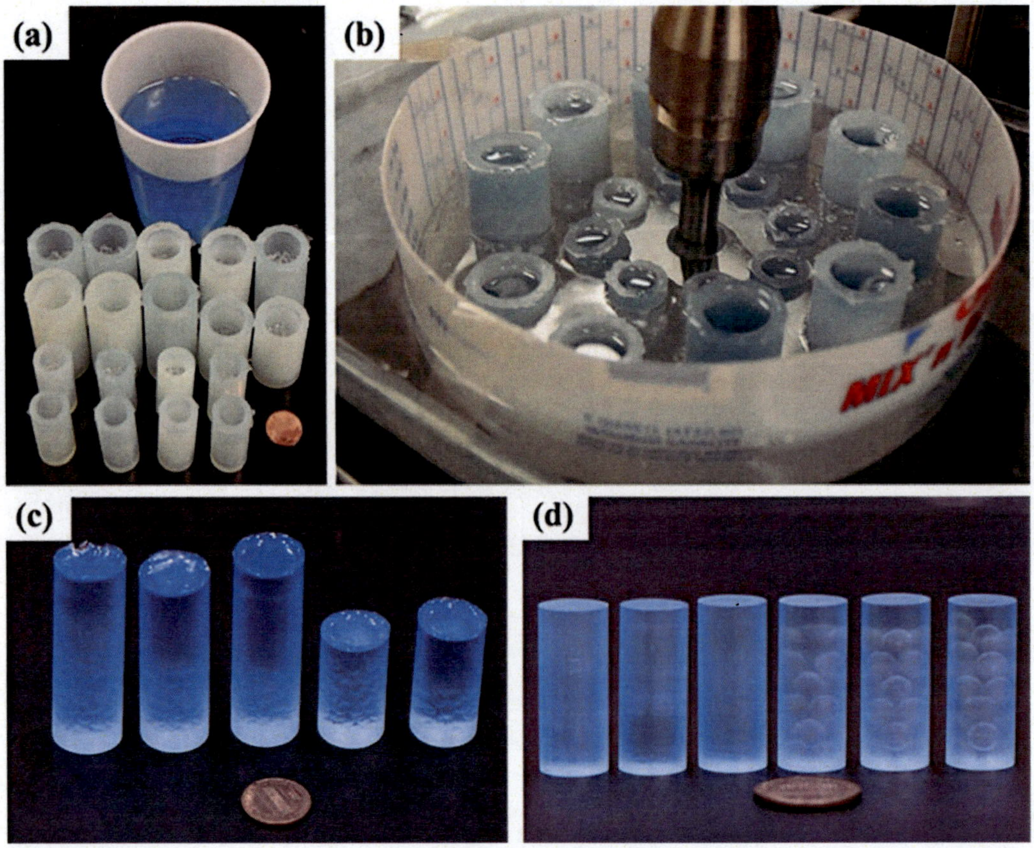

Fig. 7.1 (**a**) Premixed epoxy resin system and the silicone molds filled with cleaned and dried glass beads. (**b**) The glass bead/epoxy resin inside the customized ultrasonic water bath. The fully cured glass bead/epoxy composite sample (**c**) before and (**d**) after polishing

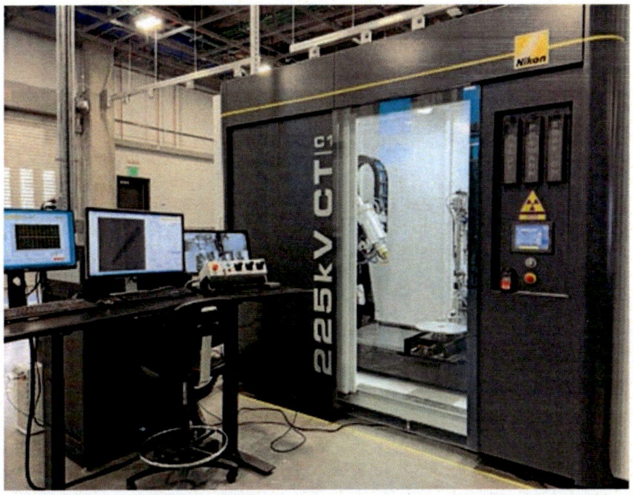

Fig. 7.2 Nikon in situ X-ray μCT system with integrated Psylotech loading frame

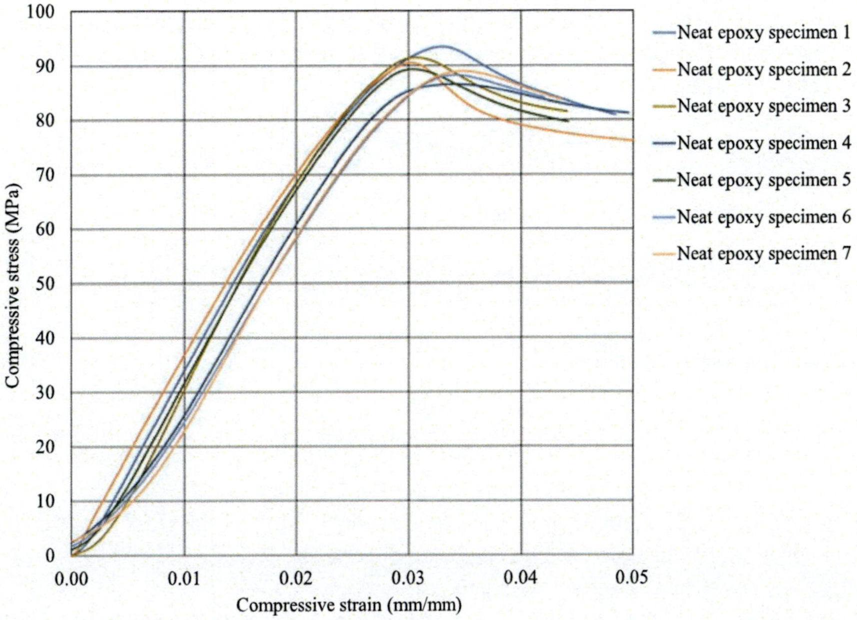

Fig. 7.3 Compressive response of the cylindrical neat epoxy specimens

with image size up to 3 μm/pixel. After reconstructing the raw μCT scan into the 3D volumetric image, a free-academic-licensed software ORS Dragonfly was used to visualize the interior of the object and inspect the incipient failure of the glass beads.

7.3 Results and Discussion

To understand the compressive behavior of the glass bead/epoxy composite, the compressive response of the neat epoxy resin was first studied. Cylindrical specimens of ½ inch and ¾ inch diameters with aspect ratios of 1:1 and 2:1 (length: diameter) were manufactured and then compressed on an Instron testing machine. The compressive stress versus strain results for all seven specimens are plotted in Fig. 7.3. And the elastic modulus is found to be 3.54 ± 0.13 GPa, and the compressive strength is calculated as 89.91 ± 2.28 MPa.

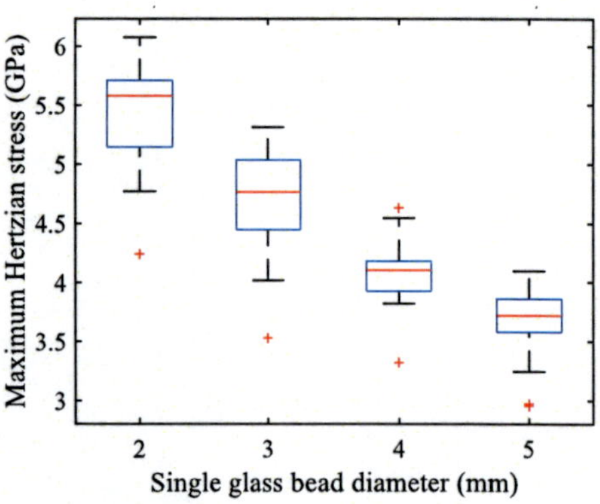

Fig. 7.4 Maximum Hertzian contact stress of single glass beads of various diameters at incipient of fracture

The glass beads are used as the inclusion in the glass bead/epoxy composite specimen, and therefore it is of interest to understand the compressive response of the single glass bead under uniaxial compression. Since glass beads have point contacts with the stainless steel confinement as well as neighboring glass beads, the Hertzian contact stress theory is applied to calculate the effective localized stress [12, 13].

When an elastic sphere of radius *R*, in this case, the glass bead, indents another elastic half-space, which is the steel compression platen, with a total deformation *d* that is recorded as the compressive displacement by the Instron machine, a contact area of radius *a* is formed in Eq. 7.1.

$$a = \sqrt{Rd} \tag{7.1}$$

Then the applied compressive force F can be calculated as Eq. 7.2 shows [13].

$$F = \frac{4}{3} E^* R^{\frac{1}{2}} d^{\frac{3}{2}} \tag{7.2}$$

where the reduced modulus $E*$ is represented in Eq. 7.3.

$$E^* = \frac{1 - v_1^2}{E_1} + \frac{1 - v_2^2}{E_2} \tag{7.3}$$

And E_1 and E_1 are the elastic modulus, and ν_1 and ν_2 are the Poisson's ratios for the glass bead and the steel compression platen, respectively. The maximum contact pressure *p* is calculated using Eq. 7.4.

$$p = \frac{3F}{2\pi a^2} \tag{7.4}$$

Therefore, the maximum Hertzian contact pressure results *p* for single glass beads of different diameters are obtained as the results are illustrated in Fig. 7.4. A decreasing trend of maximum Hertzian contact stress with respect to the increasing single glass bead diameter is observed. The maximum Hertzian contact stress for single glass beads of 2, 3, 4, and 5 mm diameter is found to be 5.42 ± 0.43, 4.69 ± 0.43, 4.07 ± 0.28, and 3.66 ± 0.31 GPa, respectively. This finding that the single glass beads of different diameters has distinct maximum failure stresses associated with the corresponding diameters is helpful when such glass beads of different diameters are used as the inclusion in the glass bead/epoxy composite specimens.

After the failure stress of the single glass bead is studied, the compressive response of the glass bead assembly of three different configurations under confinement is explored. As shown in Table 7.1, each of the three assemblies contains glass beads of the same diameters, 2 or 3 mm, and the number of glass beads in each assembly, 10, 15, and 20, respectively, is picked for the preliminary study. Figure 7.5 shows the crosshead pressure versus the compressive strain of these three glass bead assemblies. The glass bead assembly exhibits a pure elastic compressive behavior with linear compressive stress versus strain relationship only until the failure happens, at which point the load drops to zero. The slope of the linear region and the crosshead pressure at the failure of the three curves in Fig. 7.5 are calculated and summarized in Table 7.1. In comparison, the crosshead pressure at the failure for all three assemblies, 34.35, 20.42, and 7.52 MPa, respectively, are much lower than the Hertzian contact stress of the same 2 and 3 mm diameter single glass beads at the failure, which are 5.42 and 4.69 GPa,

Table 7.1 Compressive properties of glass beads assemblies of three different configurations

Assembly	Glass bead diameter (mm)	Number of glass beads (counts)	Assembly length (mm)	Assembly diameter (mm)	The slope of the linear region (GPa)	Crosshead pressure at failure (MPa)
1	3	10	10.75	6.34	4.17	34.35
2	3	15	16.12	6.34	2.55	20.42
3	2	20	6.01	6.34	2.91	7.52

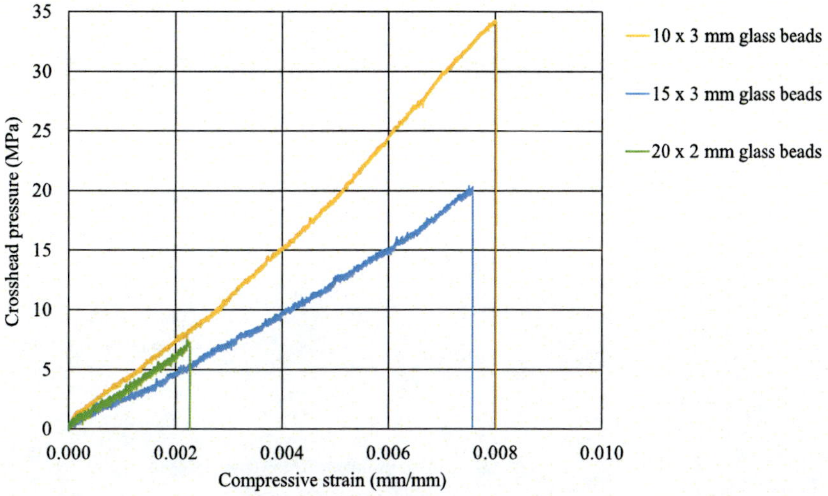

Fig. 7.5 Compressive response of the glass beads assembly under confinement

respectively. This shows the big difference between the individual glass beads and their aggregate form when it comes to load-bearing ability.

To understand the incipient failure of the glass bead assembly under confinement, X-ray μCT images of the glass bead assembly before and after the confined compression are presented in Figs. 7.6, 7.7 and 7.8. In each figure, the glass bead assembly inside the steel confining cylinder at intact state before the compression is visualized in sub-figures (a) and (b), and the assembly after failure due to the compression is shown in sub-figures (c) to (f). The compression experiment was set as that the bottom platen was moving upward to apply the compressive force, while the top platen was fixed. It can be observed from the μCT rendering that in all three configurations, glass beads toward the bottom of the assembly are crushed or damaged more than those toward the top part when the failure occurs. For the assembly consisting of ten glass beads of 3 mm diameter, three glass beads in the assembly, one of which is in contact with the bottom moving platen, and the other two are above and arranged in a chain consecutively and vertically, are crushed completely, as best visualized in Fig. 7.6c, d, while the rest of the glass beads are preserved with light to no obvious damage as shown in Fig. 7.6e, f. Similarly, in the second assembly with 15 glass beads of the same 3 mm diameter included, glass beads at the same location (most bottom and right above) are completely destroyed, whereas others incur light abrasion as shown in Fig. 7.7c–f.

However, the incipient failure is different in the assembly of 20 glass beads of 2 mm diameter as shown in Fig. 7.8. The use of smaller glass beads leads to a denser packing in the aggregate assembly, which results in a larger total contact area between the glass beads as well as between the glass beads and the sidewall of the steel confinement. No severe nor visible damage occurs on the glass beads on the outside of the assembly, but after inspecting the interior of the aggregate, one completely crushed glass bead located on the bottom center is found as revealed in Fig. 7.8e, f.

Comparing the crosshead pressure at the failure, the first assembly with ten glass beads of 3 mm diameter bears the largest pressure upon failure and also exhibits the most rigid behavior upon loading among all three assemblies in terms of compressive stress versus strain response as shown in Fig. 7.5 and Table 7.1. The significant lower pressure at the failure of the other two assemblies is most likely due to a different transmission of force-bearing mechanism of the glass bead aggregate assembly as a whole, which needs further investigations.

Next, the compressive response of the glass bead/epoxy composite assemblies under confinement is studied. Figure 7.9a, b present the compressive stress versus strain plots of the composite cylindrical specimens with 4 and 5 mm glass beads in the inclusions, respectively. The X-ray μCT scan of the intact specimen before compression was reconstructed into the

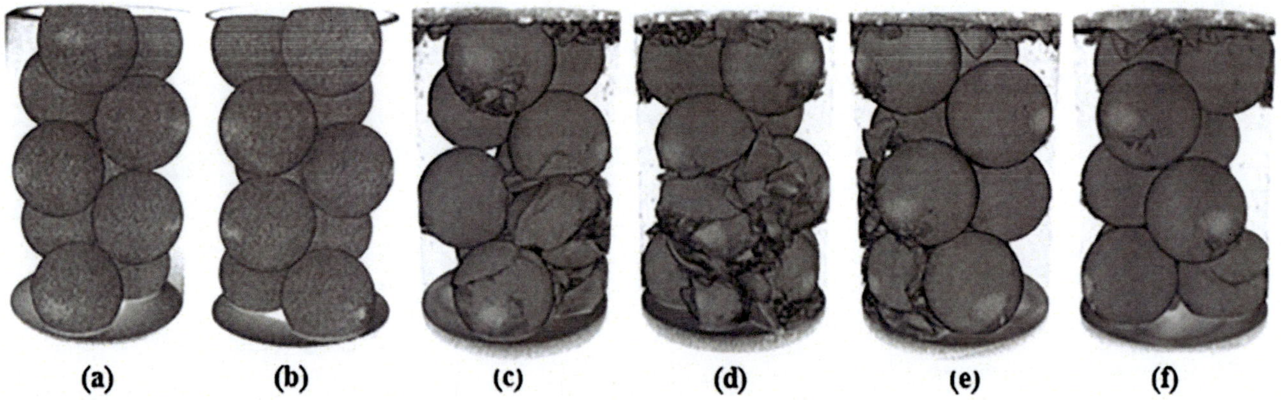

Fig. 7.6 X-ray μCT images of the ten glass beads of 3 mm diameter assembly under confinement: (**a**, **b**) at intact state before the compression and (**c–f**) at failure state after the compression

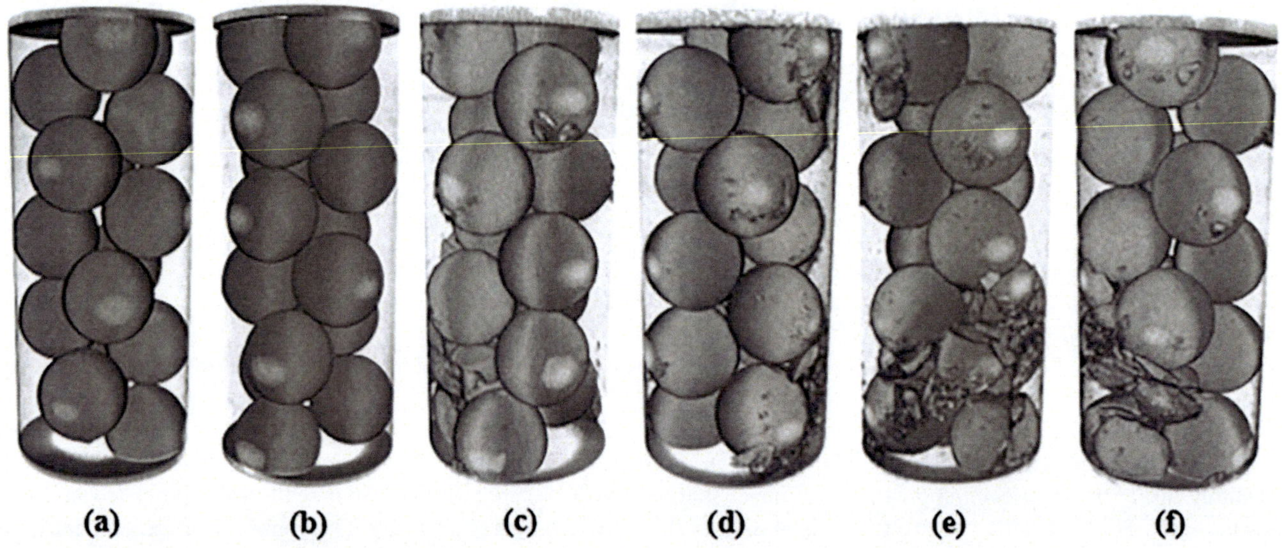

Fig. 7.7 X-ray μCT images of the 15 glass beads of 3 mm diameter assembly under confinement: (**a**, **b**) at intact state before the compression and (**c–f**) at failure state after the compression

volumetric image, and after further image segmentation, as illustrated in Fig. 7.11, the volume of the epoxy binder and the glass beads inclusions were obtained, from which the volume fraction of the inclusion was calculated. And together with the mechanical properties obtained from the loading curves, the results are included in Table 7.2. Although all specimens are prepared in the same manner and are free of defects based on the visual and μCT scan inspections, they exhibit different mechanical properties upon uniaxial compression. The most influential factor is the volume fraction of the inclusion, in this case, the glass beads in the epoxy binder. As the glass bead volume fraction increases from approximately 0.49 to 0.55, as shown in Fig. 7.10, a clear trend of increasing elastic modulus is observed from all specimens, although the same conclusion cannot be distinctively drawn for the compressive strength (Fig. 7.11).

The X-ray μCT scan captures the failure mechanism of the glass bead/epoxy composite specimen as shown in Figs. 7.12 and 7.13. Figure 7.12a, c render the volumetric image of the complete specimen at failure, including the epoxy binder and the glass bead inclusions, and Fig. 7.12b, d filter out the lower density epoxy material and thus only show the higher density glass beads in the matrix. Figure 7.13 shows three cross-sectional μCT images in the x-y, x-z, and y-z planes. It can be observed that only very few glass beads toward one end of the specimen fractured, while the rest preserved their intact form fairly well. The cracks on those fractured glass beads have the same direction as the fracture and crack opening on the epoxy binder as shown in all μCT rendering in Fig. 7.12. And this direction is vertically aligned with the two compression platens on the two ends, which indicate the location and direction of the force chain that bears the compressive load within the specimen.

The delamination occurring between the glass bead inclusions and the epoxy binder is another cause of the failure of the specimen shown in Fig. 7.14. The specimens shatter, fracture, and fail in a way of exhibiting the brittleness part of the glass

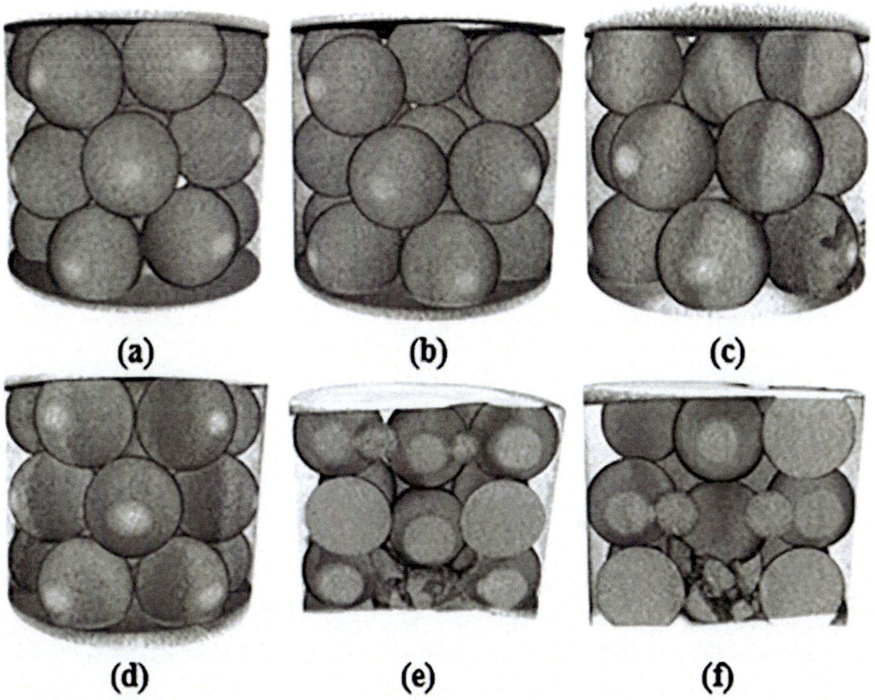

Fig. 7.8 X-ray μCT images of the 20 glass beads of 2 mm diameter assembly under confinement: (**a**, **b**) at intact state before the compression and (**c–f**) at failure state after the compression

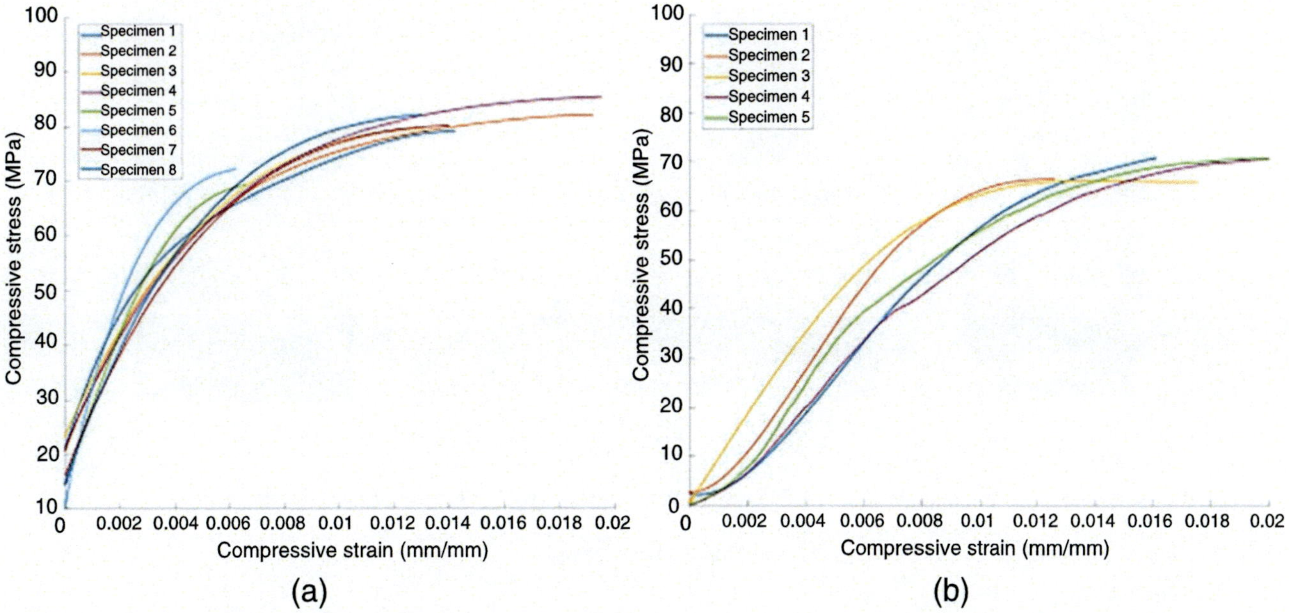

Fig. 7.9 Compressive response of the glass bead/epoxy assembly under confinement with (**a**) 4 mm and (**b**) 5 mm glass bead inclusions

beads properties rather than presenting the viscoelastic side of the compressive response of the epoxy binder of the matrix. Compared with the neat epoxy cylinders with an elastic modulus of 3.54 GPa and the compressive strength of 89.91 MPa, the glass bead/epoxy composite specimens have a much higher elastic modulus of 11.39 and 9.37 GPa, but a quite lower compressive strength of 79.55 and 70.41 MPa for both assembly configurations as shown in Table 7.2. Although the volume fraction of glass beads inclusion is about 49–55% of the entire specimen, the stiffness of the glass bead/epoxy composite specimen as a whole is greatly increased by adding the glass beads in the inclusion, represented by the much higher elastic modulus value. Meanwhile, the compressive strength is sacrificed compared to the neat epoxy specimen. From the μCT

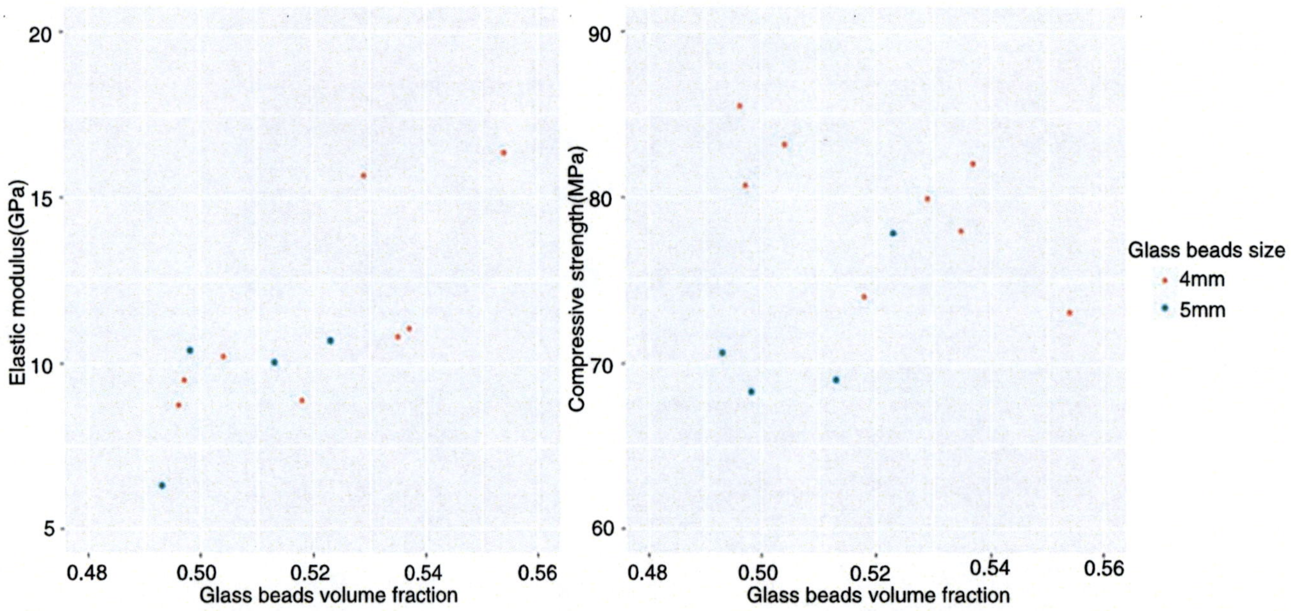

Fig. 7.10 Compressive properties of the glass bead/epoxy specimens versus the corresponding glass beads volume fraction

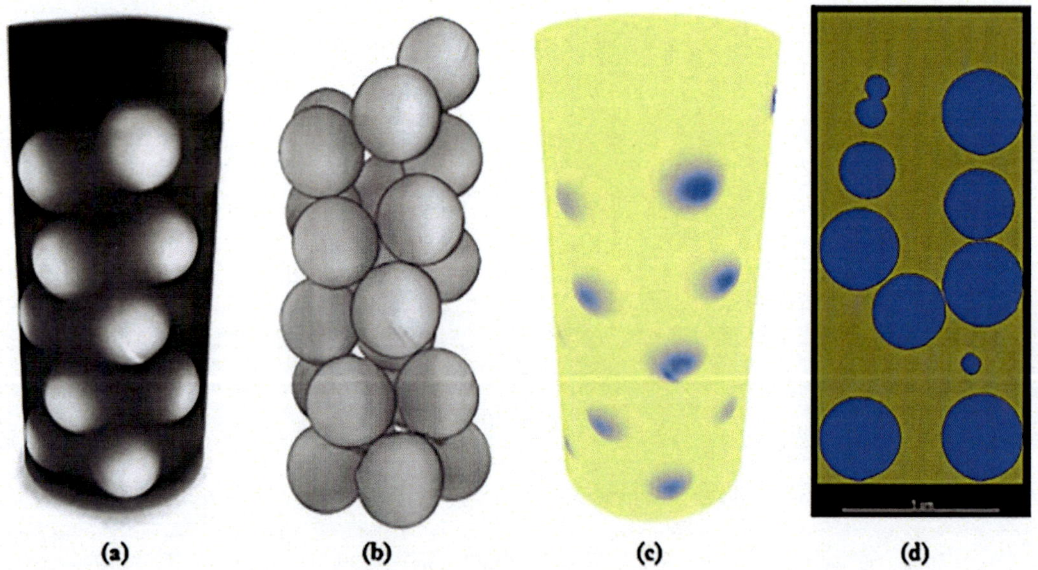

Fig. 7.11 Rendering and segmentation of the X-ray μCT images of the intact glass bead/epoxy assembly before compression

Table 7.2 Compressive properties of glass bead assemblies of three configurations

Glass bead diameters in assembly	Number of specimens tested	Elastic modulus (GPa)	Compressive strength (MPa)	Glass beads volume fraction
4 mm	8	11.39 ± 2.97	79.55 ± 4.34	0.521 ± 0.021
5 mm	5	9.37 ± 2.05	70.41 ± 4.70	0.495 ± 0.029

images, it can be deduced that the interfacial delamination between the glass beads and the epoxy blinder is the primary cause of the strength reduction.

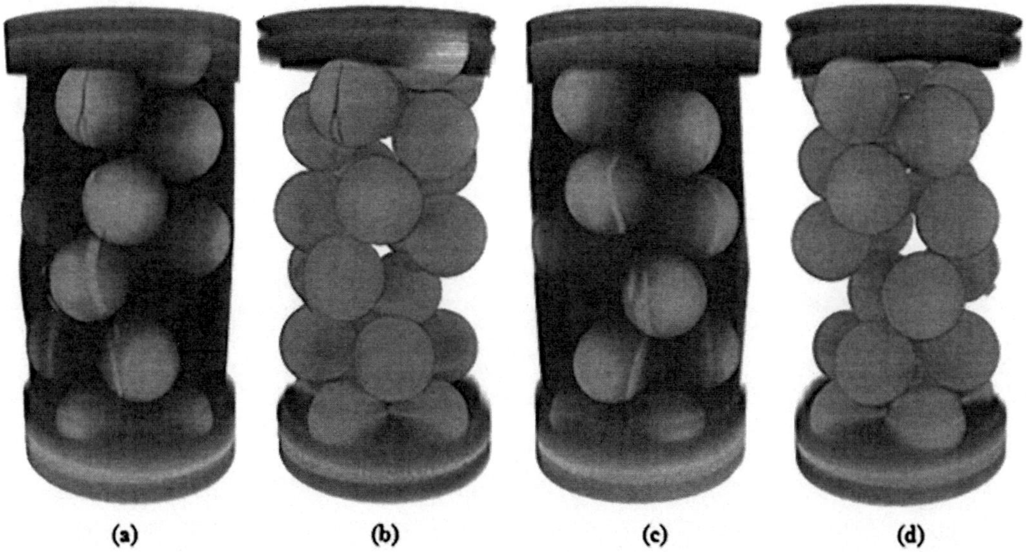

Fig. 7.12 Rendering of the X-ray μCT images of the glass bead/epoxy assembly after failure

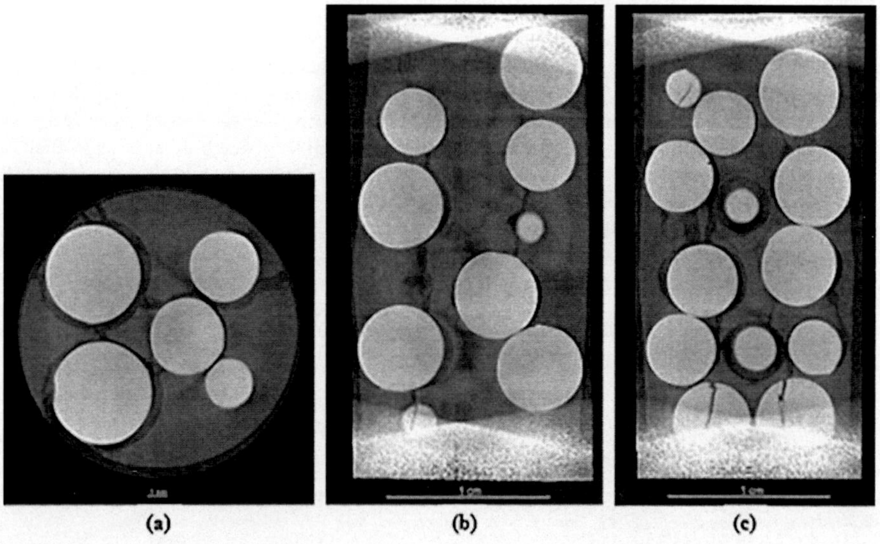

Fig. 7.13 Three cross-sectional X-ray μCT images in the x-y, x-z, and y-z planes of the glass bead/epoxy assembly after failure

7.4 Conclusion

This study focuses on understanding the incipient failure of the glass beads assembly under confined compression and the glass bead/epoxy composites under compression. The in situ X-ray μCT compression was performed, and the loading history of all the specimens was obtained. Several observations and findings are presented:

1. Single glass beads of different diameters show different compressive strength at the failure, with an increasing maximum Hertzian contact stress with respect to a decreasing sphere diameter.
2. The incipient failure of glass bead assembly under confined compression is visualized from the reconstructed μCT volumetric image. It is observed that the glass beads toward the loading end tend to be crushed with a few adjacent ones fractured, whereas most of the rest glass beads remain intact or get light abrasion or damage.
3. Clear force chains and load-bearing planes on the glass bead/epoxy composite specimens are observed. The mechanical properties of the composite specimens are greatly influenced by the volume fraction of the glass beads inclusion, and the interfacial delamination between the glass beads and the epoxy binder is likely the cause of failure upon compressive loading.

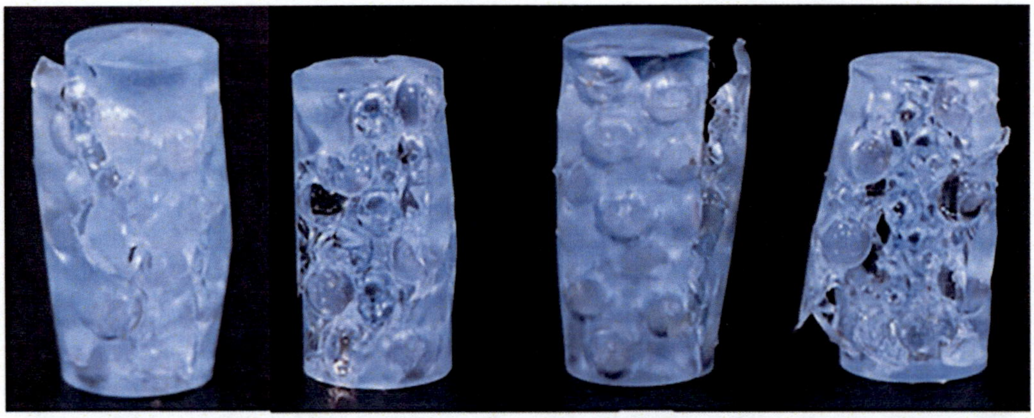

(a) (b)

Fig. 7.14 The glass bead/epoxy composite specimens after failure of (**a**) specimen 1 and (**b**) specimen 2

With future finite element analysis on this problem, more situations regarding the effects of the packing of the glass bead assembly and the glass bead/epoxy composites can be created and simulated. The incipient failure of those assemblies and glass bead/epoxy composites can be understood and further explored.

Acknowledgments The authors acknowledge support by the Department of Energy, National Nuclear Security Administration, Predictive Science Academic Alliance Program (PSAAP) under Award Number DE-NA0003962 and DE-NA-0003525. This report was prepared as an account of work sponsored by an agency of the US Government. Neither the US Government nor any agency thereof, nor any of their employees, makes any warranty, express or implied, or assumes any legal liability or responsibility for the accuracy, completeness, or usefulness of any information, apparatus, product, or process disclosed, or represents that its use would not infringe privately owned rights. Reference herein to any specific commercial product, process, or service by trade name, trademark, manufacturer, or otherwise does not necessarily constitute or imply its endorsement, recommendation, or favoring by the United States Government or any agency thereof. The views and opinions of authors expressed herein do not necessarily state or reflect those of the US Government or any agency thereof. We also acknowledge the support of NSF CMMI-1661246, CMMI-1636306, and CMMI-1726435, and the support by the Louis A. Beecherl Jr. endowed chair. We are grateful to Hexion Inc. for providing the epoxy used in this study.

References

1. Luo, H., Du, Y., Hu, Z., et al.: High-strain rate compressive behavior of glass beads under confinement. Exp. Mech. **55**, 935–950 (2015). https://doi.org/10.1007/S11340-015-9995-2/FIGURES/13
2. Hu, Z., Du, Y., Luo, H., et al.: Internal deformation measurement and force chain characterization of Mason sand under confined compression using incremental digital volume correlation. Exp. Mech. **54**, 1575–1586 (2014). https://doi.org/10.1007/S11340-014-9915-X/FIGURES/13
3. Yamamuro, J.A., Bopp, P.A., Lade, P.V.: One-dimensional compression of sands at high pressures. J. Geotech. Eng. **122**, 147–154 (1996). https://doi.org/10.1061/(ASCE)0733-9410(1996)122:2(147)
4. Luo, H., Lu, H., Cooper, W.L., Komanduri, R.: Effect of mass density on the compressive behavior of dry sand under confinement at high strain rates. Exp. Mech. **51**, 1499–1510 (2011). https://doi.org/10.1007/S11340-011-9475-2/FIGURES/13
5. Zhang, R., Chen, H., Malakooti, S., et al.: Quasi-static and dynamic confined compressive behavior of glass beads by in-situ X-ray micro-computed tomography, p. 12. ASME International Mechanical Engineering Congress and Exposition, Proceedings (IMECE) (2021). https://doi.org/10.1115/IMECE2020-23399
6. Vallejo, L.E., Lobo-Guerrero, S., Chik, Z.: A network of fractal force chains and their effect in granular materials under compression. Fractal Eng. **2**, 67–80 (2005). https://doi.org/10.1007/1-84628-048-6_5
7. Peters, J.F., Muthuswamy, M., Wibowo, J., Tordesillas, A.: Characterization of force chains in granular material. Phys. Rev. E Stat. Nonlinear Soft Matter Phys. **72**, 041307 (2005). https://doi.org/10.1103/PHYSREVE.72.041307/FIGURES/15/MEDIUM
8. Delenne, J.Y., Topin, V., Radjai, F.: Failure of cemented granular materials under simple compression: experiments and numerical simulations. Acta Mech. **205**, 9–21 (2009). https://doi.org/10.1007/S00707-009-0160-9
9. Combe, G., Roux, J.N.: Strain versus stress in a model granular material: a Devil's staircase. Phys. Rev. Lett. **85**, 3628 (2000). https://doi.org/10.1103/PhysRevLett.85.3628
10. Bardenhagen, S.G., Brackbill, J.U.: Dynamic stress bridging in granular material. J. Appl. Phys. **83**, 5732 (1998). https://doi.org/10.1063/1.367429
11. Tsoungui, O., Vallet, D., Charmet, J.C.: Numerical model of crushing of grains inside two-dimensional granular materials. Powder Technol. **105**, 190–198 (1999). https://doi.org/10.1016/S0032-5910(99)00137-0

12. Hertz, H.: Ueber die Berührung fester elastischer Körper. J. Reine Angew. Math. **1882**, 156–171 (1882). https://doi.org/10.1515/CRLL.1882.92.156/MACHINEREADABLECITATION/RIS
13. OPUS 4 | Über die Berührung fester elastischer Körper und über die Härte. http://publikationen.ub.uni-frankfurt.de/frontdoor/index/index/docId/15902. Accessed 19 Feb 2022

Chapter 8
Embrittlement of Semicrystalline Polymers: A Dynamic Fracture Analysis

J.-B. Kopp and J. Girardot

Abstract The fracture properties of polymers are complex to estimate because they strongly depend on the viscoplastic behavior of the polymer. This is especially true for semicrystalline thermoplastic polymers used for the transport or storage of fluids under pressure. The methodology presented in this work allows the construction of a part of the kinetic law of fracture of polymers, the one that reveals the brittleness of the polymer. The fracture energy of the material is estimated by taking into account the cracking regime and thus the potential inertial effects induced during the rapid propagation of cracks in the case of brittle behavior. Polyamide-11, a polymer used for the manufacture of hydrogen tank liners, is studied. A decrease in fracture energy is observed between the initiation resistance and the fast propagation resistance. This is referred to as polymer embrittlement. Electron microscopy analysis of fracture surfaces in three different zones (initiation, propagation, arrest) has revealed the mechanisms responsible for the ductility of the polymer in quasi-static regime and its brittleness in dynamic regime.

Keywords Polymer · Dynamic · Fracture · Damage · Semicrystalline · Brittle

8.1 Introduction

The fracture behavior of polymeric materials has been extensively studied in the literature. Establishing a kinetic law of fracture is convenient in the case of polymers. This one generally describes the evolution of a fracture parameter of the LEFM formalism (or of another one) as a function of the cracking velocity [1]. For most polymers, it is accepted that the fracture energy increases with the cracking velocity [1]. In other words, the faster the crack goes, the more energy it consumes. This observation is valid for amorphous thermoplastic polymers and thermosets when macromolecular mobility is limited (at intermediate velocities and temperatures). For thermoplastic polymers whose intrinsic nature or environment allows a significant macromolecular mobility revealing a potential viscoelastoplastic behavior, the tendency is not the one described above. Indeed, it has been shown for rubber toughened and semicrystalline TP polymers that the fracture energy decreases with the cracking velocity [2–4]. For these polymers, the best fracture energy of the material is expressed at intermediate velocities and temperatures. The formation of cracks, associated with cavities (damage) and fibrils (plasticity), ensures the good impact resistance of semicrystalline materials. This is why these polymers are widely used for fluid transport. Nevertheless, under specific conditions, polymer brittleness can be expressed. Rapid crack propagation appears in place of crazes. When this mechanism appears, it is possible to use the linear elastic fracture mechanics (LEFM) formalism to estimate a fracture parameter [5]. When the propagation regime is slower, visco-plasticity and volume damage mechanisms are possible for some polymers. In this case, the LEFM formalism is often not adapted anymore.

The design of industrial structures using these materials under severe stresses (low temperatures, external impact, fatigue) should consider the lower resilience of the material. This is relatively little applied because of the complexity of implementation. Revealing the brittleness of the polymer can be tricky. HDPE is an example that requires a combination of low temperature and external impact to achieve this. In this work, the strip band specimen geometry is used to study the brittleness of polyamide-11, a polymer used as a liner for hydrogen storage. The fracture energy is estimated by taking into account the dynamic structural effects induced by the rapid crack propagation. In situ analysis is limited by the duration of

J.-B. Kopp (✉) · J. Girardot
Institute of Mechanics and Mechanical Engineering (I2M), Arts et Metiers Institute of Technology, Bordeaux, Talence, France
e-mail: jean-benoit.kopp@ensam.eu

A. Amirkhizi et al. (eds.), *Challenges in Mechanics of Time-Dependent Materials & Mechanics of Biological Systems and Materials, Volume 2*, Conference Proceedings of the Society for Experimental Mechanics Series, https://doi.org/10.1007/978-3-031-17457-5_8

the test (a few hundred microseconds). The *postmortem* analysis of the fracture surfaces allows to explain the differences in fracture behavior related to the microstructure, between the initiation, rapid propagation, and arrest phases of the main crack.

8.2 Methodologies

The material studied is polyamide 11 (PA11), the Rilsan BESNO TL grade supplied by Arkema. The monomer is a castor oil derivative. It is therefore considered as a semicrystalline bio-based polymer. The degree of crystallinity of the polymer is 22%. It was measured by differential scanning calorimetry. The strip band specimens were machined from pre-injected plates. The dynamic modulus E_d was obtained by measuring the average ultrasonic velocity in the material $\langle v_u \rangle$ which is equal to 2100 ± 46 m/s. As a dynamic modulus, it can be also taken as the "relaxed" modulus of the material which avoid viscous effect in the estimation of the internal energy [2]. The value $E_d = 1620 \pm 82$ MPa is obtained from [4], with $\nu = 0.43$ and $\rho = 1040$ kg/m^3. For the estimation of the fracture energy in dynamic propagation regime, only E_d and ν will be considered to describe the behavior of the material [6].

The strip band specimen (SBS) geometry [3, 7] is used to study rapid crack propagation (RCP) and crack arrest in PA11 plates (see in Fig. 8.1). This geometry allows precise control of the boundary conditions during propagation. It generates very few inertia effects induced by rapid crack propagation. An experimental device based on these studies [2, 3, 8] has been designed.

Two notched head-to-tail plates are prestressed (see Fig. 8.1). One plate is cracked (the plate 1). The second one ensures the symmetry of the load during propagation. The specimens are initially uniaxially prestressed to a constant deformation using a universal tensile-compression machine. The loading is imposed by moving the crosshead. Once the prestress is reached, the displacement is maintained. The system is then locked by means of nuts mounted on four threaded rods. The rods pass through the jaws. The nuts stop the jaws to block the movement in the axis of the rods. The flexibility of the system is then limited. The stop of the crack is generated on the plate 1. This plate is deliberately longer than the jaw such that an unloaded zone exists. The energy release rate is almost nil and therefore very significantly lower than the fracture energy of the material. The crack is stopping at this zone, and to access this arrest zone, the plate was cut with a band saw.

To neglect viscous effect of the material, a relaxation time of the specimen (about 15 times the preload time of the specimen) is considered before crack initiation. Crack initiation is then generated by the impact of a mass on a razor blade in contact with the tip of the crack. The test is carried out at ambient temperature. Six polyamide plates were cracked. The dimension of the plates is L = 300 mm (length), H = 80 mm (height), b = 4 mm (thickness), and $l_n = 120$ mm (notch length).

The quasi-static energy release rate in mode I, noted G_{I0}, is defined as a function of the stress σ_{zz} (z is the direction of transverse crack propagation) by considering a plane stress state and the propagation of a single longitudinal crack (see Eq. 8.1).

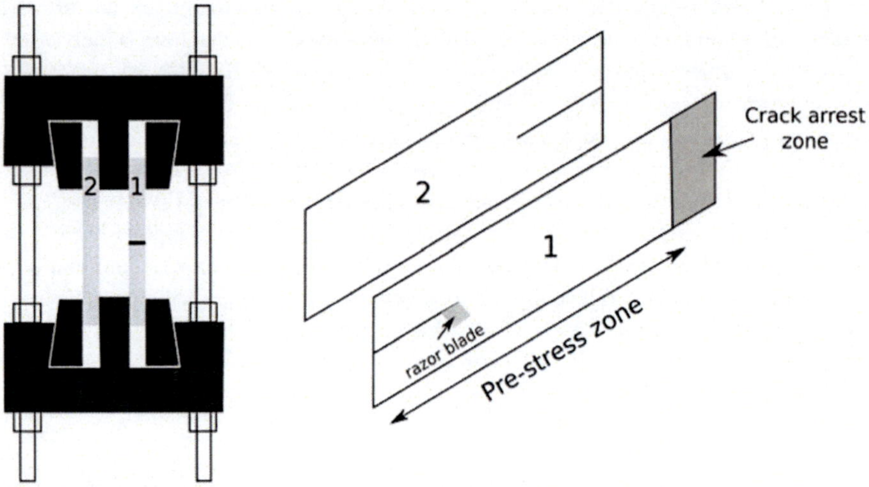

Fig. 8.1 Strip band specimen loading system

Eq. 8.1 The quasi-static energy release rate in mode I

$$G_{I0} = \frac{H\sigma_{zz}^2(1-\nu^2)}{2E}$$

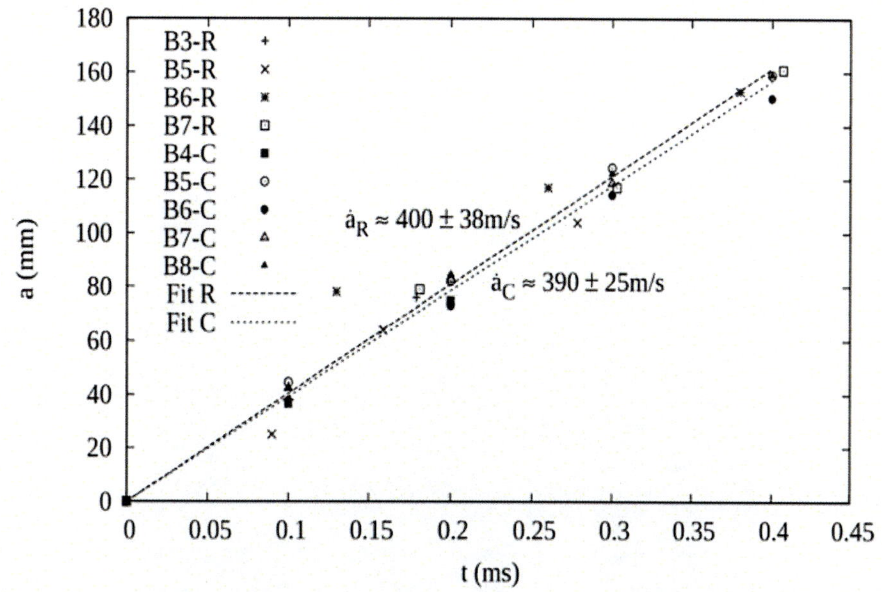

Fig. 8.2 The position of the crack tip as a function of time

A dynamic correction must be considered to estimate the energy release rate in dynamic regime, noted G_{ID}. The average crack velocity is required. A fast camera or a conductive coating [2, 9] can be used to access the spatiotemporal data of the crack tip during propagation. The latest generation of cameras has a completely satisfactory spatial and temporal resolution. The average velocity of the crack tip for this polymer is approximately 400 m/s = 0.6c_R, with c_R the Rayleigh wave speed in the material. This value was obtained by optical method and resistivity measurement (see Fig. 8.2). The conductive layer is sprayed on one side of the sample prior to the experiment. The evolution of the resistance of the R_s layer is recorded during the propagation of the fracture. An approximately constant current i is imposed with a generator G and a 10 MΩ resistor in series with the sample which has a significantly lower resistance: Rs < 45 kΩ (see in [2]). A National Instruments USB-6351 data acquisition is used to measure the potential difference u at a sampling rate of 1.25 MS/s during fracture propagation. The resistance of the conductive layer is then calculated using Ohm's law R_S = u/i. After calibration, the velocity of the crack tip at the sample scale (macroscopic velocity) is known at ±38 m/s.

The Rayleigh wave velocity (c_R) in PA11 is about 698 m/s [4]. It is known that the dynamic correction classically used for fast propagation which gives the factor $k = 1 - \dot{a}/c_R$ is unsatisfactory in the case of a SBS [10] test. A numerical model adapted to this specimen geometry has been implemented in [2] which gives a small dynamic correction of 10% which must be considered at 0.6c_R. In other words, 10% of the energy is dissipated by the structure as inertia effects when the crack propagates at 60% of the Rayleigh wave velocity in the material in the case of SBS geometry.

An optical microscope and a scanning electron microscope (SEM) are used to observe the fracture surfaces of the samples. Samples are taken from the cracked plates. They are embedded in acrylic resin and metallized in order to visualize them with the SEM. The apparatus used for metallization is a Cressington 108 auto. The gold layer is assumed to be of constant thickness on the surface of the sample. The optical microscope used is a Keyence VHX1000. The samples were then observed using a SEM (Zeiss EVOHD 15). An accelerated voltage of 10 keV is used with a current of 200 pA.

8.3 Results

The analysis of the fracture facies was performed considering three zones: initiation, rapid propagation, and crack arrest. The fracture energy value is estimated only in the propagation zone, when a quasi-permanent regime of rapid propagation is reached. Six samples were used to estimate fracture energy values (see Table 8.1). For a macroscopic velocity of the order of 0.6c_R, the fracture energy is not unique and varies from 9.4 to 15.6 kJ/m^2. This can be explained by not taking into account

Table 8.1 Estimation of the fracture energy G_{ID} (kJ/m^2), the quasi-static energy release rate G_{I0} (kJ/m^2) as a function of the crack velocity $\dot{a}$ (m/s), the temperature T (°C), and the displacement δ (mm) imposed during preloading

Sample	δ	T	$\dot{a}$	$\dot{a}/c_R$	G_{I0}	G_{ID}
B3	2	19	390	0.55	10.4	9.4
B4	2	19	420	0.60	10.1	9.1
B5	2.3	20	400	0.57	12.2	11.0
B6	2.5	21	410	0.59	15.5	14.0
B7	2.5	20	380	0.54	16.2	14.6
B8	2.7	23	390	0.55	17.3	15.6
B9	2.9	21	440	0.63	NA	NA

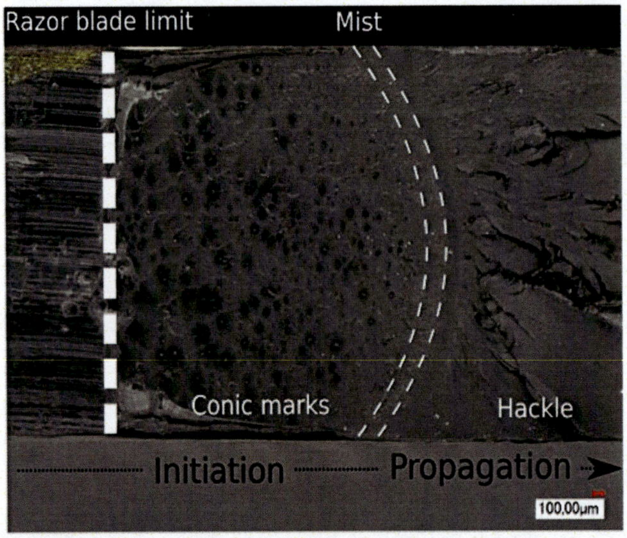

Fig. 8.3 Several zones can be identified (from left to right): the razor blade penetration zone – the conical mark zone – the intermediate "mist" zone – the chaotic "hackle" zone. The arrow indicates the direction of propagation

the amount of surface area created by the crack in the fracture energy estimate [2]. Only the amount of projected surface area (crack length multiplied by the thickness of the specimen) is considered. It is admitted that material heterogeneities can generate micro-branching of the main crack and thus surface roughness at high crack velocities [11]. This has been observed in particular for RT-PMMA [2].

For sample B9, the fracture energy was not estimated because the main crack has branched.

The initiation zone is complex. The macro-crack will indeed grow in this zone by coalescence of several micro-cracks. The energy brought to the system allowing the crack initiation is composed of the elastic energy stored in the plate and the kinetic energy generated by the impact of the mass on the razor blade. The objective of the dynamic fracture test is to establish as quickly as possible a stable state of propagation characterized by a constant energy release rate during propagation. From the point of view of crack length, this transition zone between initiation and propagation is negligible. This is the interest of a rapid propagation (i.e., dynamic fracture) test to be able to free oneself from the crack initiation. Nevertheless, on a microscopic scale, the fracture surface reflects this abrupt transition. Several characteristic areas can be observed in Fig. 8.3. From left to right, we find the zone of penetration of the blade, the zone where conical marks can be observed, an intermediate zone called "mist zone" and a chaotic zone from the point of view of roughness called "hackle zone" [12, 13].

The area of conical markings and the hackle zone are considered to be associated with crack initiation and propagation, respectively. In the mist zone (see Fig. 8.4), plastic deformation of the resin is noticeable. This suggests that the material was stressed "slowly" and at a high level of deformation.

The appearance of the conical markings in the initiation zone is well documented in the literature [14–17]. It is established that their interaction is associated with a dynamic propagation regime. Micro-cracks in this type of semicrystalline material appear to develop from the nuclei of the spherulites. Spherulites are heterogeneities (i.e., geometrical defects) which can be sites for micro-crack initiation. They are often considered more brittle than the amorphous phase of the material. The size and density of the conical marks are also indicators of the local crack velocity [18]. The higher the density and smaller the size of the conical mark, the faster the crack propagates. Here, it is observed an evolution in terms of density and size from left to right in the initiation zone (see Fig. 8.5).

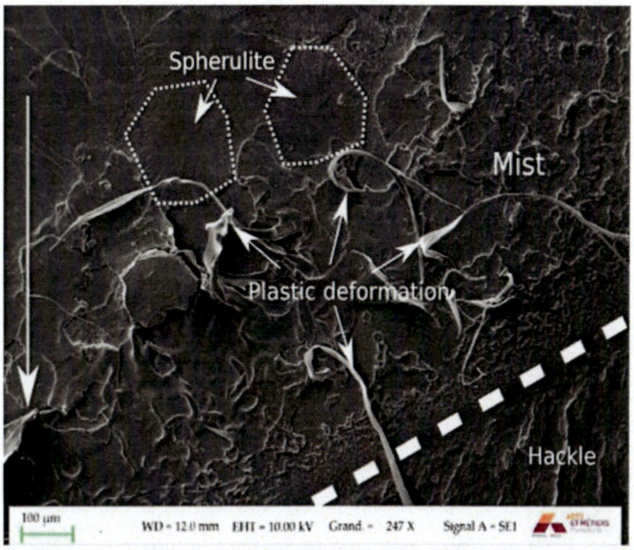

Fig. 8.4 Observation by electron microscopy of the mist zone of the fracture surface of sample B4. Several zones can be identified (from top to bottom): the smooth zone with the appearance of spherulites (end of the zone of conical marks) – the mist zone with plastic deformation of the resin – a stop mark of the macro-crack in dotted line – the hackle zone. The long arrow on the left indicates the direction of propagation

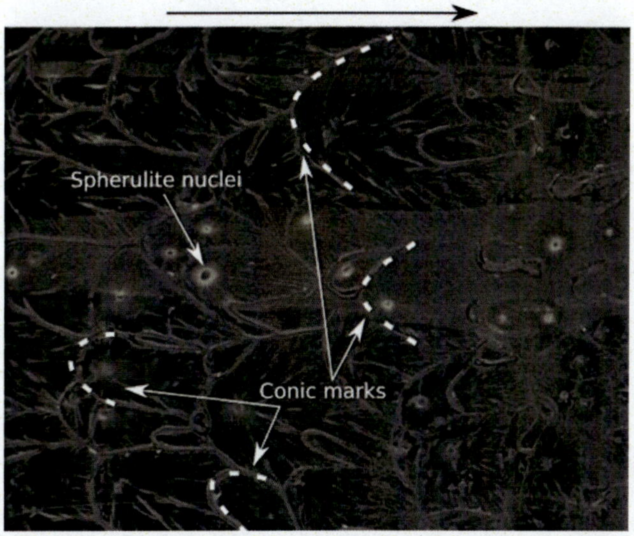

Fig. 8.5 Observation by electron microscopy of the conical marks of the initiation zone of the fracture surface of sample B3. The nuclei of the spherulites are identifiable. The arrow indicates the direction of propagation

It suggests that the macro-crack velocity was increasing very briefly until reaching the mist zone state. The area of the conical marks is known to be smooth and sometimes mirror-like on a macroscopic scale. This is confirmed by the 3D reconstruction of the fracture surface considered smooth in this area [4].

The additional energy generated by the impact of a mass on a razor blade allows the transition of the crack into a rapid propagation regime where the fracture surface is more chaotic. At this instant, the energy release rate is higher than the critical energy release rate associated with crack propagation in dynamic regime ($G_I > G_{IDc}$).

The analysis of crack propagation is relatively simple from the point of view of external loading and space-time data of the crack tip. The energy release rate is assumed to be quasi-constant over almost the entire length of the crack. The crack is in a quasi-permanent regime of rapid propagation. Analysis of the fracture surface is more complex. This makes the estimation of the fracture energy hazardous.

The zone of propagation of the fracture surface of sample B5 is shown in Fig. 8.6. The main direction of the macro-crack front is indicated by the large white arrow at the top of the figure. The three small white arrows indicate the propagation directions of several micro-cracks. Characteristic river patterns in polymer resin composites can be seen in [19].

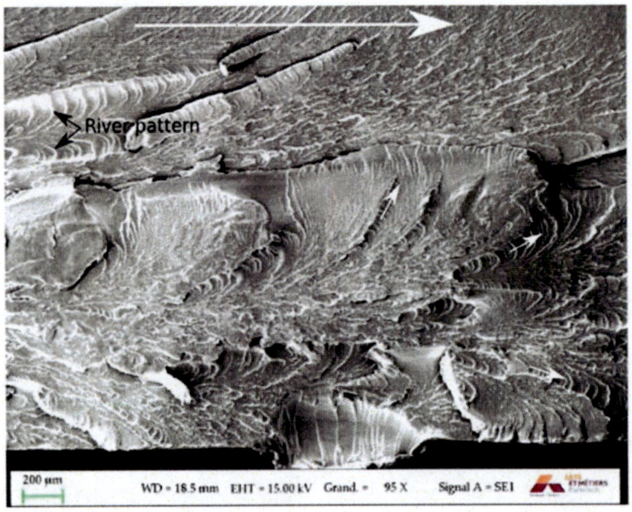

Fig. 8.6 Observation by electron microscopy of the propagation zone of the fracture surface of sample B5. The large white arrow at the top indicates the main direction of propagation of the macro-crack front. The three small white arrows indicate the propagation directions of several micro-cracks. Characteristic river patterns in multiphase polymers with high heterogeneity can be observed

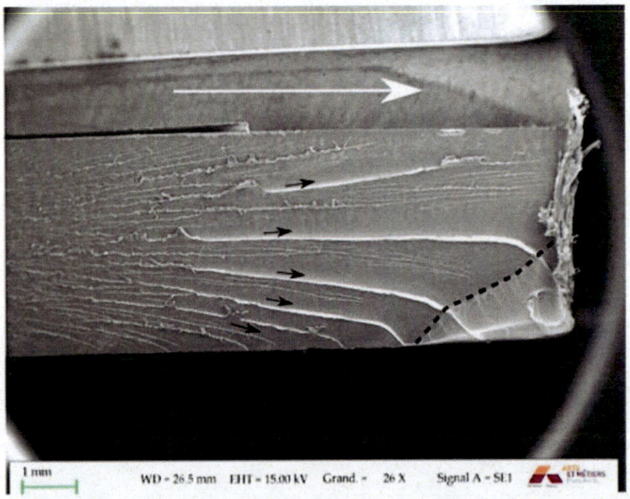

Fig. 8.7 Observation by electron microscopy of the arrest zone of a crack on the fracture surface of sample B7. The large white arrow at the top indicates the main direction of propagation of the macro-crack front. The five small black arrows indicate the directions of propagation of several micro-cracks. The dashed line indicates the arrest zone of the crack. To the right of the specimen is an area where the material has been plastically deformed as a result of cutting the specimen

The description of crack propagation through a macro-front no longer seems at all relevant at this scale. Micro-cracks seem to propagate in different directions and planes. The interaction of several micro-cracks in planes at different altitudes seems to be at the origin of the river markings. Micro-branches generating arrest marks are visible. This can be interpreted as a local stick slip phenomenon.

Figure 8.7 shows a crack arrest. The large white arrow at the top indicates the main direction of propagation of the macro-crack front. The five small black arrows indicate the propagation directions of several micro-cracks. The dashed line indicates the area of crack arrest. To the right of the specimen, there is an area where the material has been plastically deformed due to the self-heating induced by the cutting of the specimen. The micro-cracks described in the previous section continue to propagate in different directions. However, they are less and less numerous as they approach the arrest zone. The surface is smoother than during propagation. Nevertheless, cracking planes are still present at different altitudes. Crack arrest is abrupt. The macro-crack front is no longer parabolic. Conventional rib marking known for PMMA cannot be identified. Very locally one finds conical marks and spherulites as in the initiation zone. The energy release rate is no longer sufficient to bypass the spherulites. The crack comes to die in the heart of the spherulites without generating micro-branches. The evidence of these marks suggests that the local velocity is decreasing and is lower than the critical speed of micro-branching.

8.4 Conclusion

The study of fracture properties in dynamic regime was conducted for a semicrystalline thermoplastic polymer, PA11. It was found that:

- The fracture energy of the material decreases as the crack propagation velocity increases. In quasi-static regime, the value of G_{IC} is complex to estimate because it depends on the visco-plasticity of the material. It is nevertheless admitted that it is significantly higher than 15 kJ/m^2. In dynamic regime, the value is lower than 15 kJ/m^2.
- The value of the fracture energy is "not" unique at $0.6c_R$ because the real surface created by the crack is not taken into account. The surface roughness varies significantly in the dynamic regime depending on the energy release rate.
- The microscopic analysis of the fracture surfaces in three different zones—initiation, propagation, arrest—allows to understand the link between the microstructure, the local mechanisms of energy dissipation, and the fracture energy. In the dynamic regime, only the known mechanisms of heterogeneous brittle materials are observed. In the initiation and arrest phases, other dissipative mechanisms such as plastic flow are observed locally. They can explain an apparent increase in the fracture energy of the material.

To conclude, the embrittlement of semicrystalline polymers, relatively well known in industrial problems, is relatively little understood, although necessary. The methodology presented in this work allows to build a part of the kinetic law of fracture of the material, which reveals the brittleness of the polymer. Local analyses of the fracture surfaces have also allowed to understand the link between the microstructure and the cracking damage mechanisms.

Acknowledgments The authors would like to warmly thank the participation in this work of V. Honno and A. Bradu who did a Bachelor internship on the subject and J. Bega for his technical support on microscopic analyses.

References

1. Kobayashi, A.S.: Dynamic fracture toughness of Homalite-100. Exp. Mech. **18**, 11–18 (1978)
2. Kopp, J.B., Schmittbuhl, J., Noel, O., Lin, J., Fond, C.: Fluctuations of the dynamic fracture energy values related to the amount of created fracture surface. Eng. Fract. Mech. **126**, 178–189 (2014)
3. Fond, C., Schirrer, R.: Dynamic fracture surface energy and branching instabilities during rapid crack propagation in rubber toughened pmma. Notes au CRAS Ser Iib. **329**(3), 195–200 (2001)
4. Kopp, J.B., Girardot, J.: Dynamic fracture in a semicrystalline biobased polymer: an analysis of the fracture surface. Int. J. Fract. **226**, 121–132 (2020)
5. Kalthoff, J.F.: On the measurement of dynamic fracture toughnesses? A review of recent work. Int. J. Fract. **27**, 277–298 (1985)
6. Bradley, W., Cantwell, W., Kausch, H.: Viscoelastic creep crack growth: a review of fracture mechanical analyses. Mech Time Depend Mater. **1**(3), 241–268 (1997)
7. Nilsson, F.: Dynamic stress-intensity factors for finite strip problems. Int. J. Fract. **8**, 403–411 (1972)
8. Kopp, J.B., Schmittbuhl, J., Noel, O., Fond, C.: A self-affine geometrical model of dynamic rt-pmma fractures: implications for fracture energy measurements. Int. J. Fract. **193**(2), 141–152 (2015)
9. Cotterell, B.: Fracture propagation in organic glasses. Int. J. Fract. Mech. **4**(3), 209–217 (1968)
10. Broberg, K.B.: The propagation of a brittle crack. Arkiv for Fysik. **18**, 159 (1960)
11. Sharon, E., Fineberg, J.: Microbranching instability and the dynamic fracture of brittle materials. Phys Rev B Condens Matter Mater Phys. **54**(10), 7128–7139 (1996)
12. Johnson, J., Holloway, D.: On the shape and size of the fracture zones on glass fracture surfaces. Philos. Mag. **14**(130), 731–743 (1966)
13. Ravi-Chandar, K., Knauss, W.: An experimental investigation into dynamic fracture: Ii. Microstructural aspects. Int J Fract. **26**(1), 65–80 (1984)
14. Kies, J., Sullivan, A., Irwin, G.: Interpretation of fracture markings. J. Appl. Phys. **21**(7), 716–720 (1950)
15. Yang, B., Ravi-Chandar, K.: On the role of the process zone in dynamic fracture. J. Mech. Phys. Solids. **44**(12), 1955–1976 (1996)
16. Dalmas, D., Guerra, C., Scheibert, J., Bonamy, D.: Damage mechanisms in the dynamic fracture of nominally brittle polymers. Int. J. Fract. **184**(1–2), 93–111 (2013)
17. Ravi-Chandar, K., Knauss, W.: An experimental investigation into dynamic fracture: iii. On steady-state crack propagation and crack branching. Int. J. Fract. **26**(2), 141–154 (1984)
18. Guerra, C., Scheibert, J., Bonamy, D., Dalmas, D.: Understanding fast macroscale fracture from microcrack post mortem patterns. Proc. Natl. Acad. Sci. **109**(2), 390–394 (2012)
19. Barbosa, L., Bortoluzzi, D., Ancelotti, J.A.C.: Analysis of fracture toughness in mode ii and fractographic study of com-posites based on elium® 150 thermoplastic matrix. Compos. Part B Eng. **175**, 107082 (2019)

Chapter 9
Characterization of Rate-Dependent Failure Properties of Pressure-Sensitive Adhesives

Michael J. Wald, Aaron T. Hedegaard, Ryan P. Birringer, Tobias Waffenschmidt, and Nelson Goncalves Pimentel

Abstract Pressure-sensitive adhesives (PSAs) are commonly used in applications where the adhesive is required to withstand impact events, such as the consumer electronic device, aerospace, and automotive industries. Due to the viscoelastic mechanical behavior of PSAs, it is necessary to characterize these materials using strain rates and temperatures that are relevant to the intended application. This often proves challenging as the application design typically entails complex loading and deformation modes that are difficult and costly to measure directly or replicate in a controlled setting. In the present work, coupon level test methods which isolate through-thickness tension and overlap shear failure modes are used to investigate rate- and temperature-dependent failure properties of an acrylic foam tape. The time-temperature superposition principle was utilized to develop failure master curves to characterize failure properties at rates which could not be measured directly. This approach was validated by comparing master curves created using quasi-static and impact test equipment and varying test speed by approximately three orders of magnitude.

Keywords Pressure-sensitive adhesive · Viscoelastic · Failure · Time-temperature superposition · Acrylic foam tape

9.1 Introduction

Pressure-sensitive adhesives (PSAs) are used in a variety of applications where they can experience a range of strain rates from slow rates where they are expected to maintain a bond when subjected to a constant load to high strain rates common to the consumer electronic device, aerospace, and automotive industries where the PSA is required to withstand drop or impact events. Due to the viscoelastic mechanical behavior of PSAs, it is necessary to run tests at deformation rates which are relevant to the intended application. This often proves challenging for impact and drop-type loading as it is difficult to reach relevant strain rates using traditional load frames. Methods capable of measuring force and displacement during high-rate tensile and compression tests have been described previously [1–3]. These methods are often run using specialized instrumentation and can require relatively complicated data analysis.

In this work, test methods which isolate through-thickness tension and simple shear deformation and failure modes are presented. The tests methods are compatible with both instrumented drop towers for test speeds on the order of 1 m/s and standard screw-driven load frames for test speeds on the order of 10 mm/s or slower. In conjunction with rheological analysis, the time-temperature superposition principle is implemented to quantify failure properties of the adhesives at effective rates that could not be tested directly with either of the instruments described above. The viability of the time-temperature superposition method is investigated by comparing results from high-speed tests run on the drop tower and slow-speed tests run on the load frame shifted to effective rates based on the test temperature using the shift factors determined from the rheological analysis. The purpose of this work is to assess the viability of using time-temperature superposition to

M. J. Wald (✉)
Research Specialist, 3M Company, St. Paul, MN, USA
e-mail: mjwald@mmm.com

A. T. Hedegaard · R. P. Birringer
Advanced Research Specialist, 3M Company, St. Paul, MN, USA

T. Waffenschmidt
Senior CAE Specialist, 3M Deutschland GmbH, Neuss, Germany

N. G. Pimentel
Global Application Development Senior Specialist, 3M Deutschland GmbH, Neuss, Germany

A. Amirkhizi et al. (eds.), *Challenges in Mechanics of Time-Dependent Materials & Mechanics of Biological Systems and Materials, Volume 2*, Conference Proceedings of the Society for Experimental Mechanics Series, https://doi.org/10.1007/978-3-031-17457-5_9

characterize PSA failure properties at rates that cannot be tested directly using standard test equipment by controlling the temperature at which the test is run.

9.2 Background

The viscoelastic mechanical properties of PSAs are well-known and commonly characterized using rheology techniques such as shear-mode dynamic mechanical analysis (DMA) master curves [4, 5] to quantify the rate dependency of these properties. The horizontal shift factors used to create the master curves can often be described by the Williams-Landel-Ferry (WLF) model [5] which has the following form:

$$\log(a_T) = \frac{-C_1 (T - T_{\mathrm{ref}})}{C_2 + (T - T_{\mathrm{ref}})} \tag{9.1}$$

where C_1 and C_2 are fitting constants, a_T is the shift factor for temperature T, and T_{ref} is the chosen reference temperature. This model assumes the material is linear viscoelastic and deformations are small.

Like the bulk mechanical properties, PSA failure is also a rate-dependent phenomenon [6]. Despite the PSA undergoing large deformations, Kaelble demonstrated that the WLF relationship from a shear rheology master curve could be used to create a master curve to describe the relationship between effective peel rate and steady-state peel force [7]. In addition, time-temperature superposition has been used several times to quantify rate-dependent failure properties of polymeric materials including PSAs and epoxies [6, 8–12].

For homogeneous, isotropic linear viscoelastic materials, energy dissipated during failure of a viscoelastic polymer is a combination of the surface energy required to create two new surfaces and the dissipation of elastic strain energy in the bulk due to viscoelasticity [13]. This concept is expressed using the functional form:

$$W = W_0 (1 + \varphi (v, T)) \tag{9.2}$$

where W is the total energy dissipated, W_0 is the critical energy required for the material to fail in the limiting case where the test speed is slow, and the test temperature is high, minimizing viscous effects. $\phi(v,T)$ is a dimensionless function of the testing rate and testing temperature representing viscous dissipation. For polymer systems that follow the time-temperature superposition principle, $\phi(v,T) = \phi(a_T{*}v)$, where a_T is the shift factor used to scale the test rates run at a given temperature relative to effective test rates at the reference temperature T_{ref}. A relationship between $\phi(a_T{*}v)$ and loss tangent, also known as tan(δ), has been proposed previously based on strain energy release rate measurements using a probe tack test [14]. In addition, Xu and Dillard [15] described a relationship between tan(δ) and Mode I fracture toughness, indicating a correlation between energy dissipation due to bulk viscoelasticity and the apparent energy necessary to initiate fracture or failure in a viscoelastic polymer.

9.3 Experimental Setup

Experiments were completed for an acrylic foam tape with a nominal thickness of 1.1 mm. To complete this analysis, shear mode rheology, through-thickness tension, and double-overlap shear tests were completed. The experimental procedures for each of these tests are described below.

Shear Mode Rheology

A small strain shear-mode master curve was obtained as follows: an 8 mm diameter disk was punched from a 1.1 mm thick sheet of the adhesive. The liners were removed from the adhesive disk, and the disk was then loaded between 8 mm diameter parallel plates on an ARES-G2 strain-controlled rotational rheometer (TA Instruments, New Castle, DE, USA). Oscillatory shear frequency sweeps were conducted at temperatures from 150 °C to −80 °C in 5 °C increments. At each temperature step, the sample was subjected to oscillatory shear deformations with an initial strain amplitude of 5%, with frequencies ranging from 0.1 Hz to 50 Hz. Auto-strain was used to automatically adjust the strain amplitude downward to limit the torque on the rheometer to 20 g-cm. This ensured that the measurements were conducted in the linear viscoelastic range of the material, regardless of the current testing temperature. The resulting frequency sweeps were shifted onto a master curve using time-temperature superposition (TTS) with a reference temperature of 25 °C, optimizing for superposition of the shear

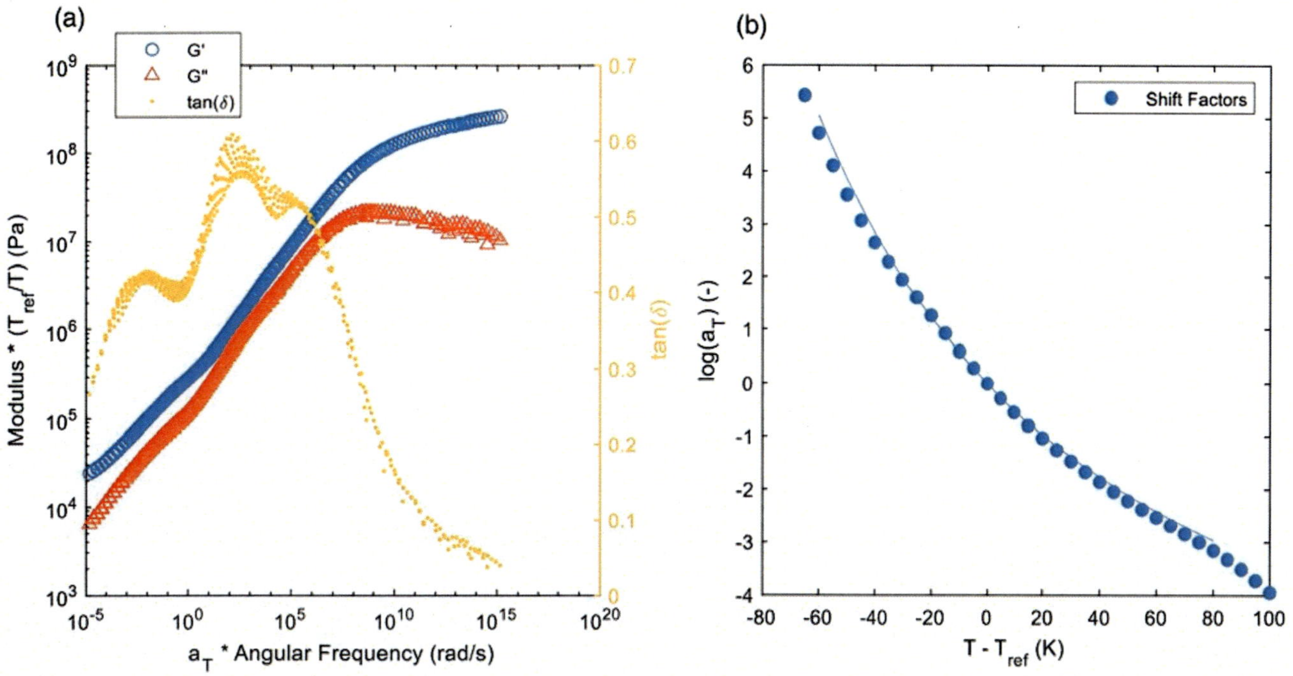

Fig. 9.1 (**a**) Storage modulus, G′; loss modulus, G′′; and loss tangent, tan(δ); as a function of effective angular frequency and (**b**) associated shift factors as a function of relative temperature along with the fitted WLF model (solid line)

storage modulus (G'). Prior to TTS, a baseline y-axis shift was applied to the dynamic modulus terms (shear storage modulus G' and shear loss modulus G"), according to the following equation:

$$b_T = \frac{T_{ref}}{T} \tag{9.3}$$

where b_T is the vertical shift factor by which the dynamic moduli at temperature T (in Kelvin) were multiplied before shifting and T_{ref} (in Kelvin) was the reference temperature of 298.15 K. This vertical shifting was performed to account for the entropic origin of polymer elasticity and hence improves the overall superposition of the dynamic moduli. Figure 9.1 shows the resulting master curves and the corresponding shift factors, a_T. The WLF relationship (Eq. 9.1) was fit to the data where a_T is the shift factor by which the frequencies (or rates) at temperature T are multiplied to create the master curve at the reference temperature, T_{ref}, of 25 °C. The fit parameters were $C_1 = 9.27$ and $C_2 = 170.2$ K. As shown in Fig. 9.1b, the WLF relationship fits the data well for all temperatures tested.

Through-Thickness Tension

Tensile impact specimens were made by creating a construction of two stainless steel plates bonded with a layer of the acrylic foam tape. The test specimen geometry is shown in Fig. 9.2. Before applying the adhesive, the faces of the blocks were cleaned with isopropyl alcohol. After the alcohol evaporated, 3M Primer 94 (3M Company, St Paul, MN, USA) was applied. After the primer dried, the specimens were assembled. The assembled specimens were individually compressed for 30 s with a 4 kg weight. The specimens were then conditioned at 80 °C for 30 min. Following the heat treatment, the specimens were conditioned at 23 °C 50% RH for 48 h.

Drop Tower Test Method Tensile impact tests were run using an Instron CEAST 3490 drop tower with crosshead 7510.021 and striker 7519.696 (Instron Corporation, Norwood, MA, USA). A 22 mm diameter, 5 mm thick striker tip was used. Specimens were constrained with a pneumatic clamp. Force and acceleration were measured with a +/− 22 kN piezoelectric load cell. The striker mass was 13 kg. Tests were run at temperatures between −30 °C and 100 °C. A drop height of 100 mm was used for all tests, corresponding to an impact velocity of approximately 1.4 m/s. Five replicates were run for each temperature and rate permutation.

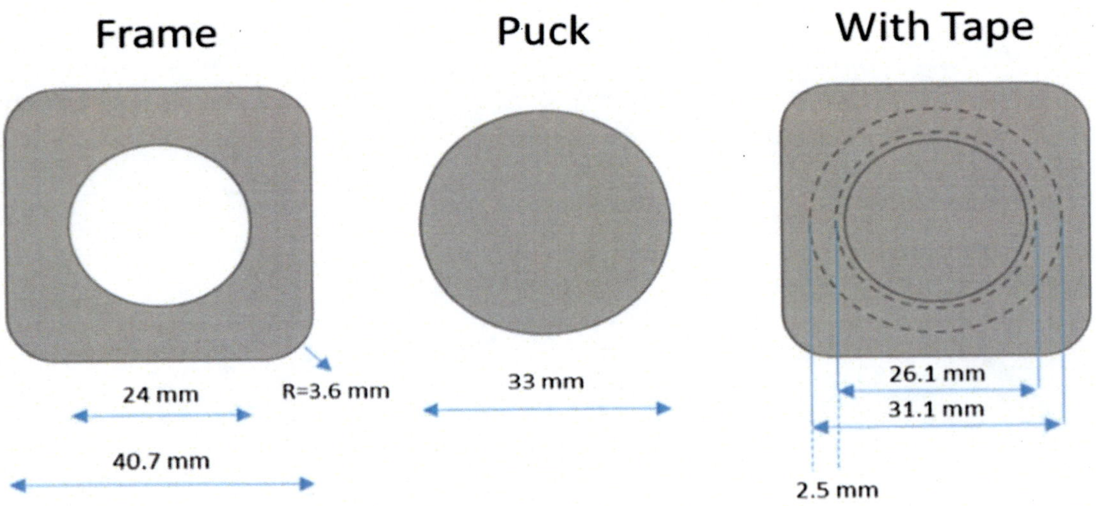

Fig. 9.2 Geometry of the two substrate pieces used to make the through-thickness tension specimen. When assembled, the striker passes through the frame substrate and strikes the puck substrate. The bonded area is 225 mm^2

Fig. 9.3 Schematic of the double-overlap shear specimen used for the drop tower tests

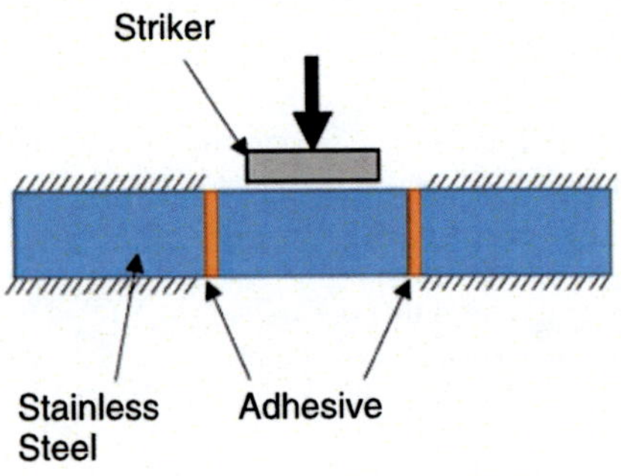

Quasi-Static Test Method Specimens for quasi-static tensile tests were made using the method described above. The fixture used to mount the specimens in the drop tower was adapted for an electromechanical load frame. Tests were completed on a Criterion Model 43 (MTS Corporation, Eden Prairie, MN, USA) with a +/− 10 kN load cell. A 19 mm diameter aluminum cylindrical flat punch served as the striker. Tests were run at rates between 0.1 mm/s and 8 mm/s and temperatures between −100 °C and 100 °C. Five replicates were run for each temperature and rate permutation.

Double-Overlap Shear

Shear impact specimens were made of three stainless steel blocks bonded with two pieces of the adhesive (Fig. 9.3). Similarly, quasi-static specimens were made by bonding two stainless steel posts with a single layer of adhesive. The 10 mm × 10 mm block faces were bonded with 10 mm × 10 mm of adhesive. Before applying the adhesive, the faces of the blocks were cleaned with methyl ethyl ketone and primed using 3M Primer 94. The assembled specimens were individually compressed for 30 s with a 4 kg weight. The specimens were then conditioned at 80 °C for 30 min. Following the heat treatment, the specimens were conditioned at 23 °C and 50% RH for 48 h.

Drop Tower Test Method Shear impact tests were run on using the setup and test parameters described above for through thickness tension with the exception that a 15 mm × 15 mm × 5 mm stainless steel striker tip was used.

Quasi-Static Test Method Shear tests were run on a Criterion Model 43 universal test frame (MTS Corporation, Eden Prairie, MN, USA) using an MTS Acumen double-lap shear test fixture that was adapted for use in the Criterion load frame. Force was measured using a +/− 10 kN load cell. Tests were run at 1 mm/s, and test temperatures were between −80 °C and 100 °C. Five replicates were run for each temperature.

9.4 Results

Peak stress, peak energy density, total energy density, and critical strain were calculated for all tests. Peak stress is the maximum nominal stress. Critical strain is the engineering strain at which peak stress occurred. Peak energy density is the integral of force with respect to displacement normalized for adhesive volume from initial contact between the striker and specimen up to the displacement at which maximum force was achieved. Total energy density is the integral of force with respect to displacement normalized for adhesive volume from the point of initial contact until the end of the test. Following Kaelble's approach [7], the strain rates at which the tests were run were shifted according to the WLF relationship for acrylic foam tape (Fig. 9.1b). Peak stress, peak energy density, total energy density, and critical strain were also corrected using the vertical shift factor (Eq. 9.3). The resulting master curves are shown for though-thickness tension and double-overlap shear in Figs. 9.4 and 9.5, respectively. For through-thickness tension, the master curves generated using the drop tower and quasi-static test method agree for the four metrics considered, indicating that time-temperature superposition is applicable. For double-overlap shear, peak stress and total energy density master curves agreed for the two approaches, but critical strain and peak energy density differed. The apparent shift between the drop tower and quasi-static data for critical strain may be due to differences in determining the initial contact for the drop tower and quasi-static test methods due to modifying the fixturing between the two instruments. The calculation for peak energy density is based on the critical strain value determined for each test. Therefore, discrepancies in critical strain between the two test methods are likely propagating through the analysis as discrepancies in peak energy density.

9.5 Discussion

Despite the tests described above being run at rates differing by approximately three orders of magnitude, the time-temperature superposition principle appears to be a valid tool for understanding the rate-dependent failure properties of the acrylic foam tape. While both instruments are capable of creating master curves, there are several aspects of the quasi-static test methods that are advantageous to the drop tower method. From a practical standpoint, load frames are more commonly available than drop towers, making the test method accessible without the need to purchase specialized high speed test equipment. In addition, master curves created using the quasi-static method covered a greater range of effective rates (Figs. 9.4 and 9.5). The drop tower tests were run from −30 °C to 100 °C due to the limits of the instrument's environmental chamber. In contrast, the chamber used for the quasi-static tests can reach −115 °C. For the acrylic foam tape considered in this work, this is sufficiently cold to compensate for the slower test speeds, extending the range of the master curve beyond the limits of the drop tower method.

The quasi-static test method also offers the potential for improved force resolution. High-frequency oscillations were often observed in the measured force from the drop tower tests. An example of this behavior is shown in Fig. 9.6a for a test run at 1.4 m/s at a temperature of 60 °C. The force oscillations are believed to be related to the system dynamics of the instrument that are being excited at the point of impact and are forces that are not necessarily experienced by the adhesive. For comparison, the force response from a quasi-static test run at 1 mm/s and 0 °C is also shown in Fig. 9.6a. Using WLF shift factors to normalize time to $T_{ref} = 25$ °C, these tests are expected to be similar, although not quite identical. The effective strain rate for the quasi-static tests was slightly faster than that of the drop tower tests, resulting in the test occurring over a shorter period as show in Fig. 9.6a. However, the shapes of the curves are roughly the same, and because the quasi-static test was run at a slow speed, the effects of system dynamics observed in the drop tower are not observed in the test data. This simplifies data post-processing and provides a clearer picture of the forces experienced by the adhesive.

The improved force measurements at this effective strain rate (~20 1/s) yield additional information that was not present in the drop tower data. In Fig. 9.4b, d, the quasi-static tests at this rate have a very large error bar for critical strain and peak energy density. When examining the force-time response in Fig. 9.6a, it appears that in this range of strain rates, there are two local maxima in the force response that are approximately equal in magnitude. The first is located at approximately 0.025 s, and the second is located at approximately 0.15 s. While peak force and total energy density are very similar for all replicates, the strain at peak force and the peak energy density have a large amount of scatter because the strain at peak force is switching between the two local maxima depending on which peak is greater in each test run. This behavior is not present at other rates because the second peak is dominant at lower effective strain rates, and the first peak is dominant at higher effective strain rates (Fig. 9.6b). This suggests a potential transition in failure mechanisms is occurring near an effective strain rate of 20 1/s, leading to the large scatter in the data. This material behavior was not evident in the drop tower data because the appearance of the local maxima was overwhelmed by high-frequency noise.

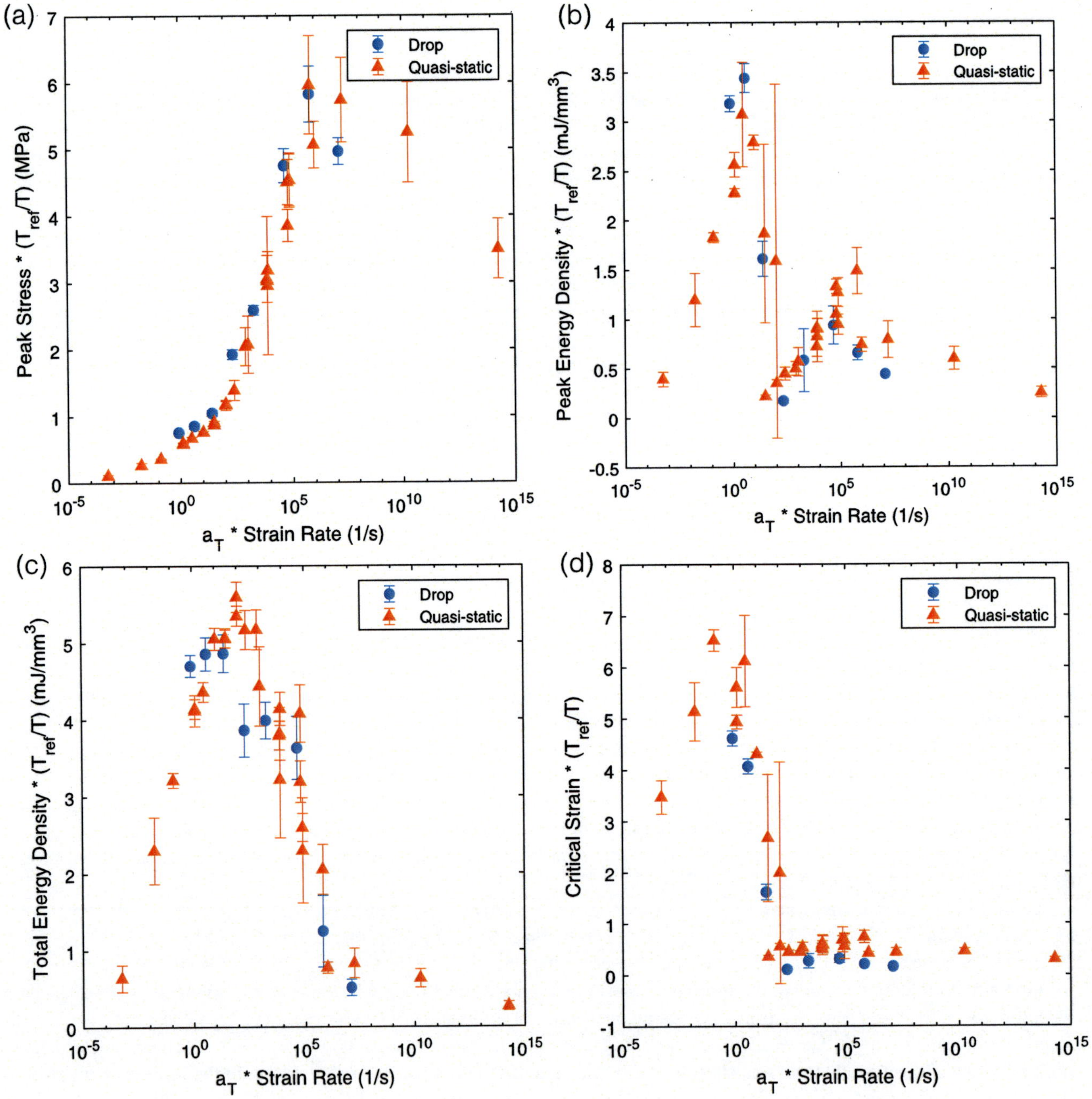

Fig. 9.4 **(a)** Peak stress, **(b)** peak energy density, **(c)** total energy density, and **(d)** critical strain as a function of effective strain rate for through-thickness tension tests run on a drop tower and a load frame

Peak energy density and total energy density master curves from the quasi-static through-thickness tension and double-overlap shear tests were compared to the tan(δ) master curve from the shear rheology experiment to investigate the correlation between failure energy and tan(δ) described in Eq. 9.2. For through-thickness tension tests (Fig. 9.7a), it was observed that the peak of the tan(δ) master curve occurred at an effective angular velocity that was similar to the effective strain rate at which maximum total energy density was reached. While both effective strain rate and effective angular velocity are plotted on the same axis, they are similar but not equivalent and are potentially the source of the horizontal offset observed between the energy master curves and tan(δ) master curve. The shapes of the two master curves are also similar suggesting that for this acrylic foam tape, small strain viscoelasticity is the primary mechanism for rate-dependent energy dissipation during failure. This was unexpected as the material underwent large strains prior to failure. The peak of the tan(δ) master curve also correlated with the local minimum observed in the peak energy density master curve around 100 1/s. This is in the same

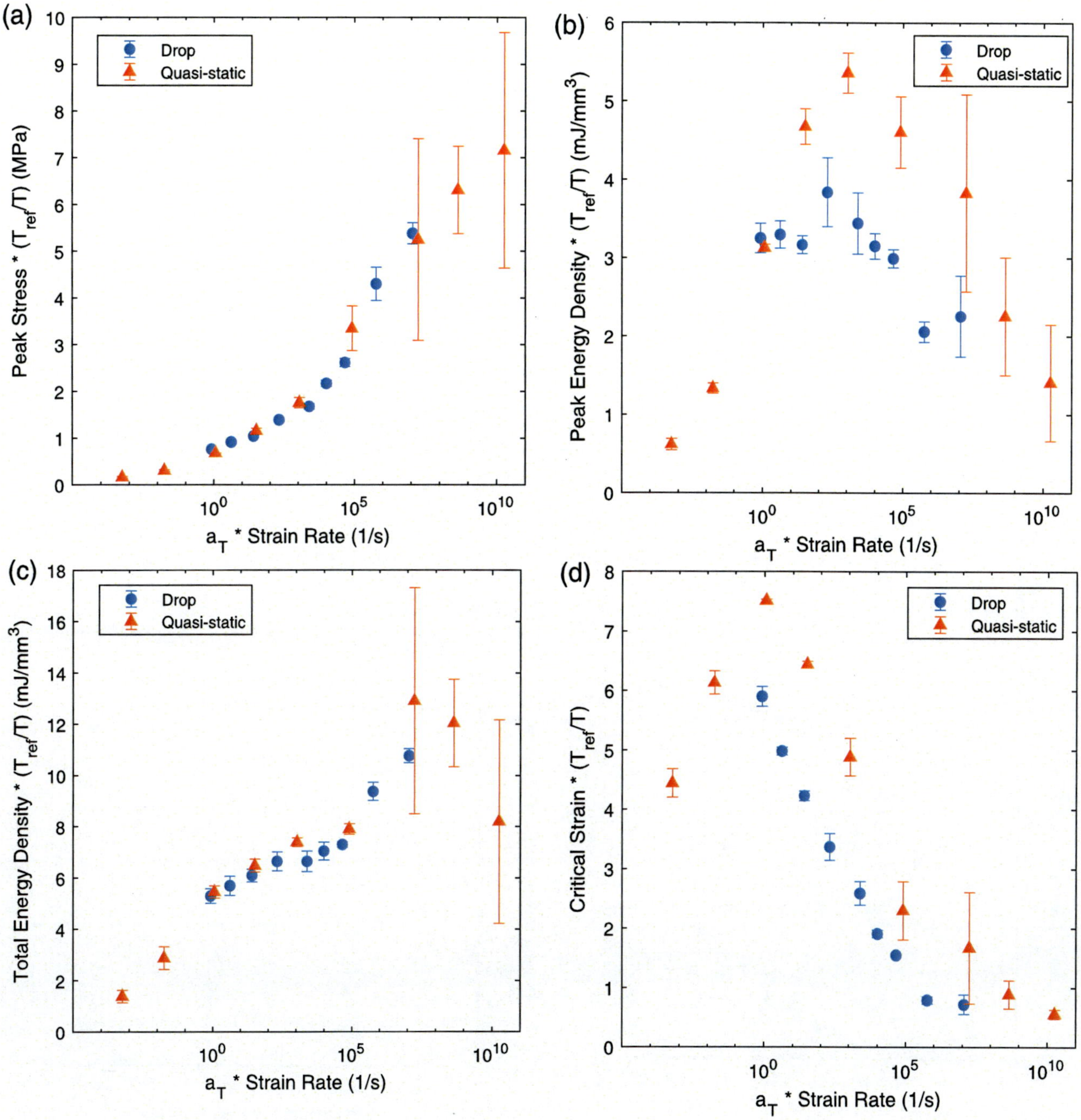

Fig. 9.5 (**a**) Peak stress, (**b**) peak energy density, (**c**) total energy density, and (**d**) critical strain as a function of effective strain rate for overlap shear tests run on a drop tower and a load frame

regime as the failure mechanism transition described earlier, suggesting that the change in failure mechanism may be related to the glass transition of the material.

Double-overlap shear peak energy density and total energy density master curves are compared with the tan(δ) master curve in Fig. 9.7b. In this case, the maximum of the peak energy density master curve correlated with the tan(δ) master curve. However, unlike the through-thickness tension master curve, the total energy density master curve for double-overlap shear does not correlate with tan(δ). This is under further investigation and may indicate that unlike through-thickness tension, large strain effects are present, and small strain viscoelasticity is not the primary mechanism for rate-dependent energy dissipation.

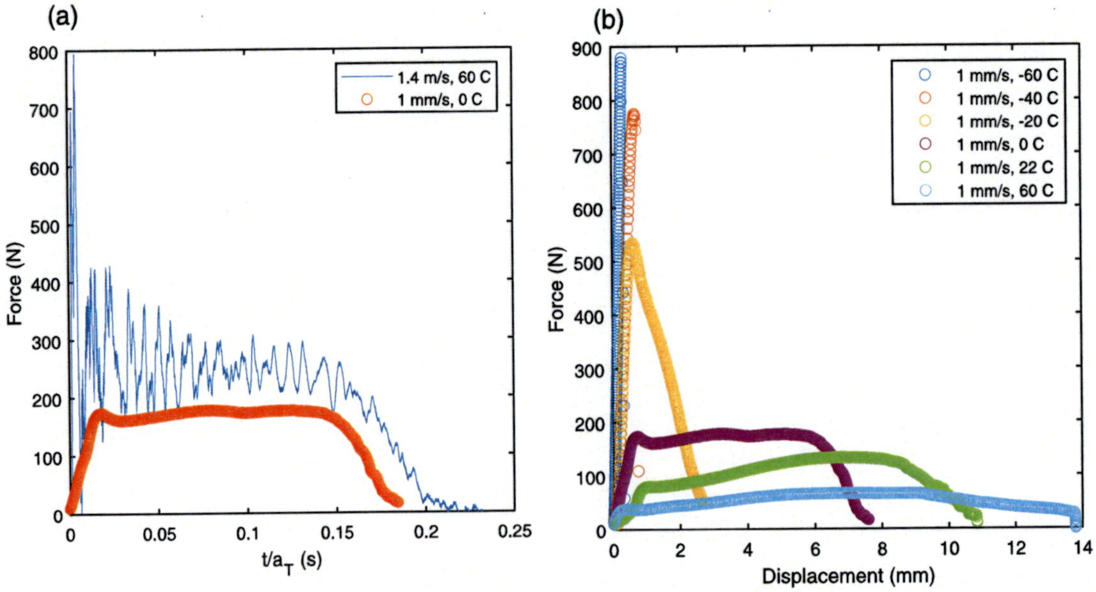

Fig. 9.6 (**a**) Comparison of drop tower and quasi-static tests run at an effective strain rate of approximately 20 1/s. (**b**) Comparison of quasi-static tests run at different temperatures showing the transition of the displacement at which peak force occurs

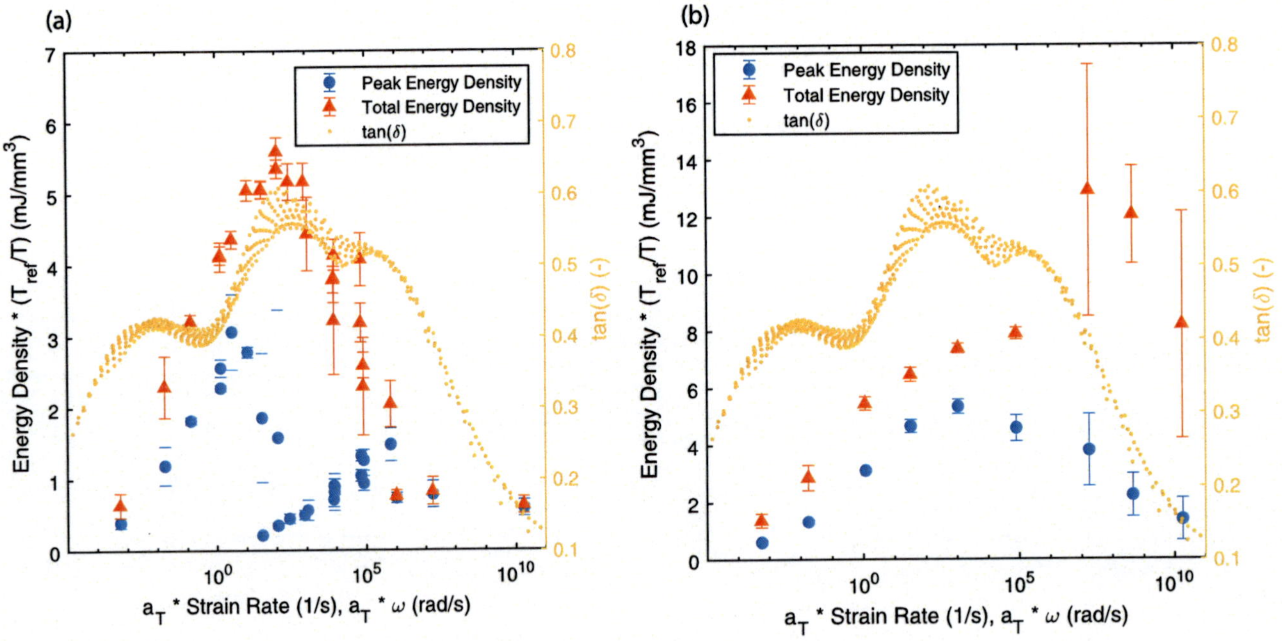

Fig. 9.7 Comparison of peak and total energy density master curves from the quasi-static test method and tan(δ) for (**a**) through-thickness tension and (**b**) double-overlap shear

9.6 Conclusion

Rate and temperature dependence of failure properties for an acrylic foam tape were investigated. Using shear rheology, through-thickness tension, and double-overlap shear tests, it was demonstrated that time-temperature superposition can be used to characterize failure properties of PSAs. This is advantageous because test temperature can be used to test at effective strain rates that the available test equipment is unable to reach. This avoids the need for specialized high-rate testing equipment. It was also demonstrated that by using slower speed tests along with lower temperatures, the force resolution could be improved over equivalent tests run approximately three orders of magnitude faster on a drop tower. A correlation between the energy density master curves and tan(δ) was also observed for through-thickness tension tests. The relationship between double-overlap shear energy density master curves and tan(δ) will be investigated further in the future.

References

1. Chen, W.W.: Experimental methods for characterizing dynamic response of soft materials. J. Dyn. Behav. Mater. **2**(1), 2–14 (2016)
2. Yokoyama, T., Nakai, K.: Determination of the impact tensile strength of structural adhesive butt joints with a modified split Hopkinson bar. Int J Adhes & Adhes. **56**, 13–23 (2015)
3. Carpenter, S.: Effects of sample preparation method on adhesive tensile impact test. 6th WCARP & Adhesion Society 41st Annual Meeting (2017)
4. Ferry, J.D.: Viscoelastic properties of polymers. In: Wiley, 3rd edn, New York (1980)
5. Williams, M.L., Landel, R.F., Ferry, J.D.: The temperature dependence of relaxation mechanisms in amorphous polymers and other glass-forming liquids. J. Am. Chem. Soc. **77**(14), 3701–3707 (1955)
6. Creton, C.: "Material science of pressure-sensitive adhesives," chapter in "Materials science and technology". Wiley-VCH Verlag GmbH & Co. KgaA (2006)
7. Kaelble, D.H.: Theory and analysis of peel adhesion: rate-temperature dependence of viscoelastic interlayers. J. Colloid Sci. **19**(5), 413–424 (1964)
8. Gent, A.N.: Adhesion of viscoelastic materials to rigid substrates. II. Tensile strength of adhesive joints. J Polymer Sci: Part A-2. **9**(2), 283–294 (1971)
9. Lim, W.W., Mizumachi, H.: Fracture toughness of adhesive joints. III. Temperature and rate dependencies of mode II fracture toughness and adhesive shear strength. J. Appl. Polym. Sci. **63**(7), 835–841 (1997)
10. Huang, Y., Kinloch, A.J.: The use of time-temperature superposition in studying the fracture properties of rubber-toughened epoxy polymers. J. Adhes. **41**(1–4), 5–22 (1993)
11. Pohlit, D.J., Dillard, D.A., Jacob, G.C., Starbuck, J.M.: Evaluating the rate-dependent fracture toughness of an automotive adhesive. J. Adhes. **84**(2), 143–163 (2008)
12. Yarusso, D.J.: Quantifying the relationship between peel and rheology for pressure sensitive adhesives. J. Adhes. **70**(3–4), 299–320 (1999)
13. Gent, A.N.: Adhesion and strength of viscoelastic solids. Is there a relationship between adhesion and bulk properties? Langmuir. **12**(19), 4492–4496 (1996)
14. Creton, C., Ciccotti, M.: Fracture and adhesion of soft materials: a review. Rep. Prog. Phys. **79**(4), 046601 (2016)
15. Xu, S., Dillard, D.A.: Determining the impact resistance of electrically conductive adhesives using a falling wedge test. IEEE Trans Compon Packag Technol. **26**(3) (2003)

Chapter 10
Evaluation of Johnson-Cook Failure Model for Aluminium Alloy AA6063-T6

Sanjay Kumar, Anoop Kumar Pandouria, Purnashis Chakraborty, and Vikrant Tiwari

Abstract In this research, the aluminium alloy AA6063-T6 was investigated for evaluation of its failure parameters at room and high temperatures. The weight percentage of different elements in AA6063-T6 was also obtained using spectroscopy. The quasi-static tests at different strain rates from room temperature to higher temperatures were performed on a universal testing machine. Notched tensile specimens of notch radii 1 mm, 2 mm, and 3 mm were used to find the effect of stress triaxialities. The tensile tests at high strain rates are performed using a tensile Hopkinson pressure bar setup. The different temperatures considered during quasi-static conditions were 25 °C, 50 °C, 100 °C, 150 °C, and 200 °C. It was found that flow stresses were increased with strain rates, whereas flow stresses were decreased at higher temperatures. Using experimental results, the Johnson-Cook failure model parameters were evaluated. The Johnson-Cook failure model parameters are used as input parameters for finite element simulation.

Keywords AA6063-T6 · Spectroscopy · Johnson-Cook Cook failure parameters · Triaxialities

10.1 Introduction

Aluminium and its alloys are widely used for structural applications due to their high strength-to-weight ratio, corrosion resistance to many chemicals, ease to machines, and nonmagnetic characteristics. Aluminium alloy of AA6xxx series is magnesium and silicon-based alloy. Magnesium is used to increase its strength, whereas silicon reduces its melting temperature [1]. The heat treatment (T6) of aluminium alloy of grade AA6063-T6 represents a solution heat treated with artificially aged. Due to the lightweight aluminium alloy, it is preferred in automobiles and light trucks. AA6063-T6 has good weldability properties, so it is used in vehicle chassis. Aluminium alloys are used in vehicles so that the weight may be saved up to 47%, and it also saves fuels as well as produces less pollution. Aluminium alloys are easily recyclable [2].

It has been seen through kinds of literature that many authors obtained Johnson-Cook plasticity and Johnson-Cook failures model parameters for different alloys of aluminium, but few are available related to the evaluation of dynamic properties of AA6063-T6. Johnson et al. [3], Lesueur [4], and Teng [5] obtained the constitutive model of AA2024-T351, whereas P. Sharma et al. [6] obtained the constitutive model and failure model parameters of AA2014-T652. Giglio et al. [7] and Dabboussi et al. [8] evaluated parameters of the constitutive model of AA6061-T6. Fracture model parameters of different aluminium alloys AA5083-H116 and AA7075-T651 were obtained by authors Clausen et al. [9] and Brar et al. [10], respectively.

S. Kumar (✉)
Delhi Technological University, Delhi, India

A. K. Pandouria · P. Chakraborty · V. Tiwari
Indian Institute of Technology, Delhi, New Delhi, India
e-mail: purnashis.chakraborty@am.iitd.ac.in; Vikrant.Tiwari@am.iitd.ac.in

A. Amirkhizi et al. (eds.), *Challenges in Mechanics of Time-Dependent Materials & Mechanics of Biological Systems and Materials, Volume 2*, Conference Proceedings of the Society for Experimental Mechanics Series, https://doi.org/10.1007/978-3-031-17457-5_10

10.2 Johnson and Cook Cumulative Damage Model

Johnson and Cook [11] suggested a cumulative damage model in order to express the fracture behaviour of materials under different loading conditions. To show fracture behaviour, a fracture parameter known as damage parameter denoted by D is defined as the summation of the ratio of equivalent plastic strain increment $\Delta\varepsilon$ to equivalent fracture strain $\varepsilon_f$as mentioned in Eq. 10.1.

$$D = \sum \frac{\Delta\varepsilon}{\varepsilon_f} \tag{10.1}$$

The fracture will occur in materials if the damage parameter becomes unity. Fracture strain ε_f is expressed in terms of strain triaxiality, strain rate, and temperature, as shown in Eq. 10.2.

$$\varepsilon_f = \left[D_1 + D_2 \text{ exp } \left(D_3\sigma^*\right)\right]\left[1 + D_4 \ln \frac{\dot{\varepsilon}}{\dot{\varepsilon}_0}\right]\left[1 + D_5T^*\right] \tag{10.2}$$

where D_1, D_2, D_3, D_4, and D_5 are constants determined from experiments. The right side of the above equation has three different brackets, and each has a different role. The first bracket of the above equation relates to the effect of stress triaxiality (σ^*) and fracture strain. The second bracket represents the effect of strain rate ($\dot{\varepsilon}$), while the third bracket shows the effect of temperature (T^*) on fracture strain. T^* is the homologous temperature $\left(T^* = \frac{T-T_0}{T_m-T_0}\right)$; T_o is the room temperature, while T_m is the melting temperature of the material.

The stress triaxiality (σ^*) is the ratio of hydrostatic to the deviatoric component of stress. The expression for stress triaxiality ratio of hydrostatic (σ_h) to a deviatoric component of stress (σ_{eq}) is represented with Eq. 10.3. The hydrostatic and deviatoric stresses are further expressed in terms of the three-dimensional state of stresses.

$$\sigma^* = \frac{\sigma_h}{\sigma_{eq}} = \frac{\left(\sigma_x + \sigma_y + \sigma_z\right)/3}{\sqrt{\sigma_x{}^2 + \sigma_y{}^2 + \sigma_z{}^2 - \sigma_x\sigma_y - \sigma_y\sigma_z - \sigma_y\sigma_z + 3\left(\tau_{xy}^2 + \tau_{yz}^2 + \tau_{xz}^2\right)}} \tag{10.3}$$

The expression for stress triaxiality (σ^*) at the centre of the notch is expressed in terms of minimum diameter (a) of rod and notch radius(R) [12], as shown in Eq. 10.4. Fracture strain (ε_f) is evaluated after measurements of fractured diameter (d_f) and gauge diameter (d_0) using Eq. 10.5. Terms A_0 and A_f mentioned in Eq. 10.5 represent the original and final cross-sectional areas of the specimen before and after experiments.

$$\sigma^* = 1 + \ln\left(1 + \frac{a}{2R}\right) \tag{10.4}$$

$$\varepsilon_f = \ln\left(\frac{A_0}{A_f}\right) = 2\ln\left(\frac{d_0}{d_f}\right) \tag{10.5}$$

10.3 Experimental Procedures

10.3.1 Material

Aluminium alloy AA6063-T6 is an Al-Si-Mg alloy. The weight percentage of different elements was obtained with spectroscopy, as shown in Table 10.1. Specimens used for tensile tests for quasi-static and high strain rates tests were obtained from a circular bar of 13 mm diameter using a lathe machine.

Table 10.1 The chemical composition of AA6063-T6

Elements	Al	Cr	Cu	Fe	Mg	Mn	Si	Ti	Zn	Sn
% weight (wt%)	98.71	0.0016	0.0038	0.163	0.5476	0.0139	0.504	0.0213	0.0087	<0.001

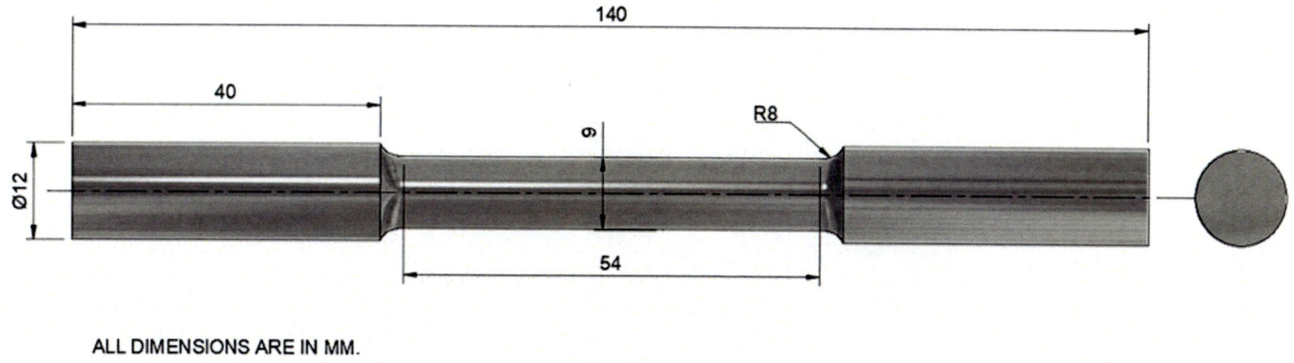

Fig. 10.1 Dimensions used for tensile test specimens of AA6063-T6

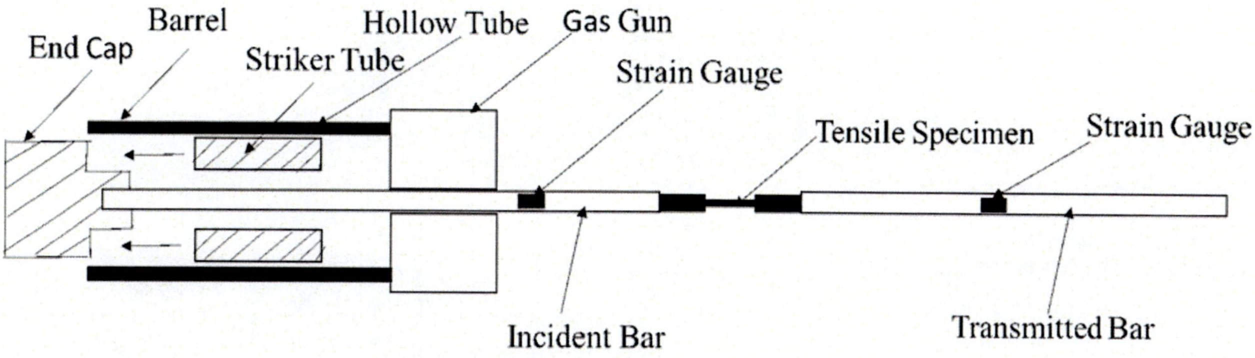

Fig. 10.2 Schematic of a tensile split Hopkinson pressure bar

10.3.2 Mechanical Testing

The quasi-static experiments were performed on a universal testing machine of 50 kN capacity model H 50 KS, whereas high strain rate tests were performed on tensile split Hopkinson bar (TSHB). The specimen with detailed dimensions used for the tensile test in the quasi-static condition is shown in Fig. 10.1. The schematic diagram of TSHB is shown in Fig. 10.2, which consists of hollow striker tube, end cap, incident, and transmitted bar. When the striker impacts the end cap, the tensile pulses are induced in the bar and propagated toward the specimen. When the propagated tensile waves pass through the specimen, the tensile tests at high strain rates are conducted at high strain rates. The strain-stress properties at high strain rates are obtained using measured reflected and transmitted tensile strain on the incident and transmitted bar, respectively.

10.4 Experimental Results

For calibration of the Johnson-Cook failure model, notched specimens of different notch radii 3 mm, 2 mm, and 1 mm having different stress triaxialities are used during experiments. Notched specimens of different notch radii with their triaxialities are shown in Fig. 10.3. The values of minimum diameter, measured at neck region of specimens, are 3 mm, 5 mm, and 7 mm for notch radius 3 mm, 2 mm, and 1 mm, respectively. Triaxialities for different notch radii of specimens shown in Fig. 10.3 are evaluated after substituting the notch radius and minimum diameter of the neck region in Eq. 10.4. The fracture strain of specimens is measured by measuring the diameter of the fractured specimen and original gauge diameter area using Eq. 10.5.

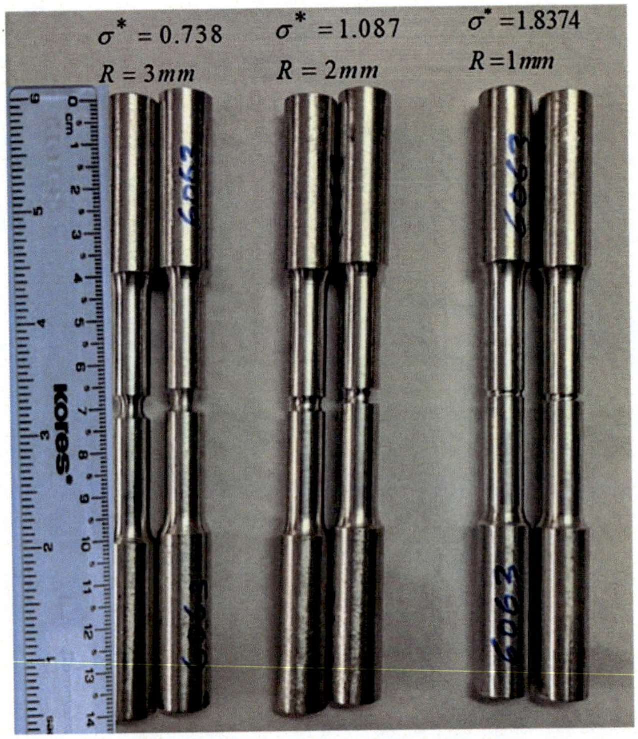

Fig. 10.3 Notched specimens with stress triaxilities

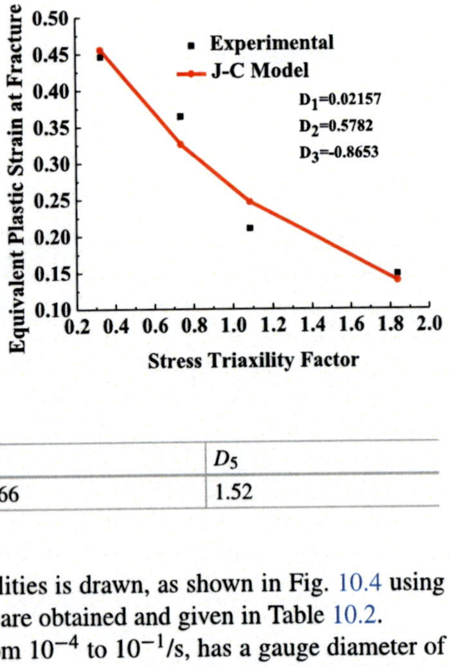

Fig. 10.4 Determination of failure parameters D_1, D_2, and D_3

Table 10.2 Johnson-Cook failure model parameter for AA6063-T6

D_1	D_2	D_3	D_4	D_5
0.02157	0.5782	−0.8653	0.066	1.52

After getting the fracture strains, the graph between fractured strains and triaxialities is drawn, as shown in Fig. 10.4 using MATLAB software, and the values of the J-C failure parameters D_1, D_2, and D_3 are obtained and given in Table 10.2.

The specimen shown in Fig. 10.1, used for quasi-static tests at strain rates from 10^{-4} to 10^{-1}/s, has a gauge diameter of 9 mm, whereas the lengths of the reduced section and gauge length measured are 54 mm and 45 mm, respectively. The total length of the specimen, including gripping length (40 mm at each side), is 140 mm, while the portion used for gripping has 12 mm diameter according to standard E8M.

Figure 10.5 shows the true stress-strain curve for AA6063-T6 at different strain rates. The quasi-static experiments performed at strain rates 10^{-4}, 10^{-3}, 10^{-2}, and 10^{-1}/s are shown in Fig. 10.5a–d, respectively, and corresponding yield stresses obtained with 0.2% offset methods are 271.85, 278.17, 282.37, and 294.46 MPa. Tensile test at strain rate 410/s is performed on tensile Hopkinson pressure bar setup shown in Fig. 10.2, and the yield strength, which is measured using stress-strain curve shown in Fig. 10.5e, is 346.07MPa. Photographs of fractured specimens at different strain rates after experiments are shown in Fig. 10.6.

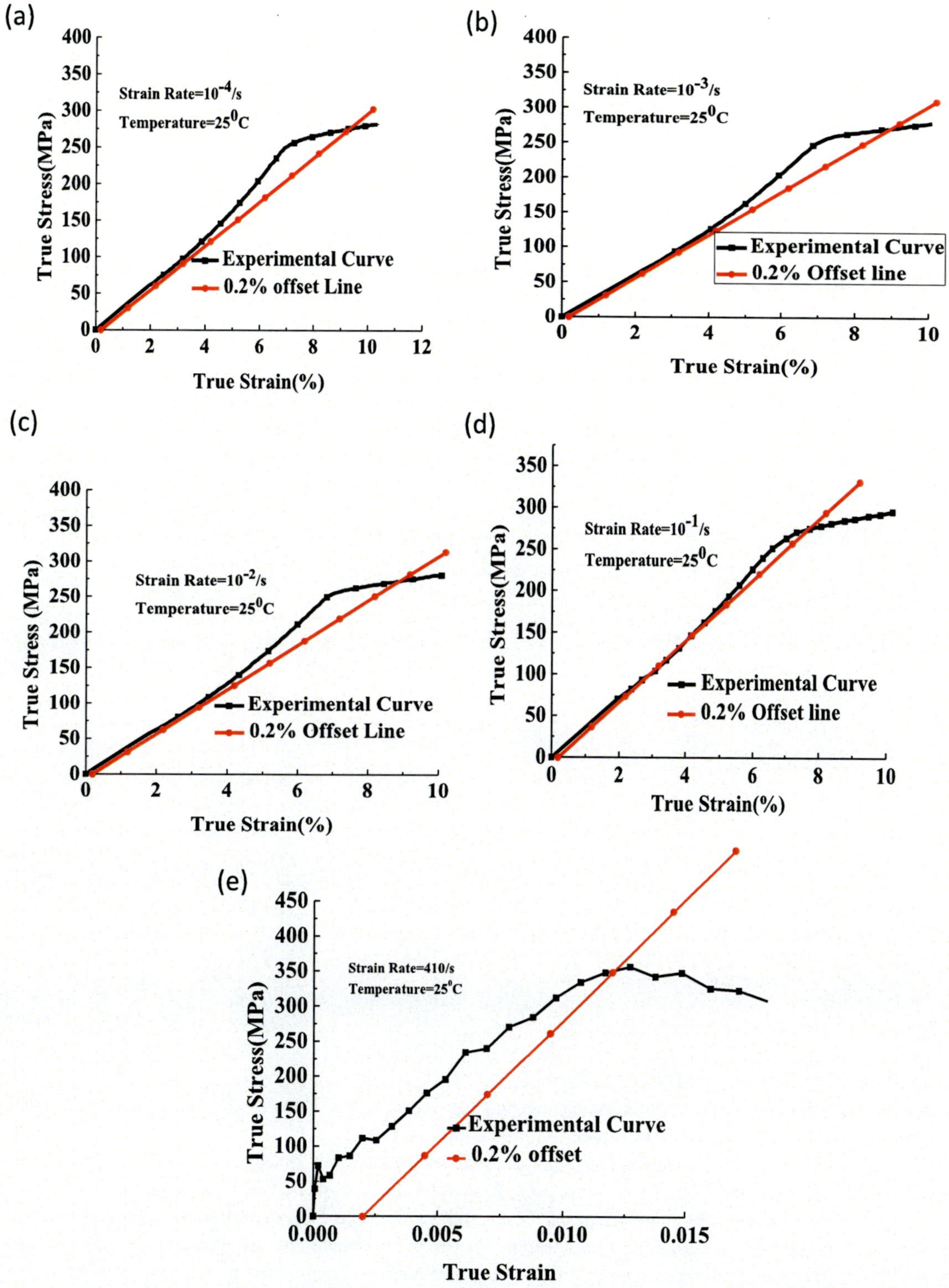

Fig. 10.5 True stress-strain curve for AA6063-T6 at different strain rates (25 °C temp) **(a)** 10^{-4}/s **(b)** 10^{-3}/s **(c)** 10^{-2}/s **(d)** 10^{-1}/s **(e)** 410/s

The strain hardening failure parameter D4 for AA6063-T4 is obtained by experiments performed on tensile specimens shown in Fig. 10.1 at different strain rates. Strains at different strain rates are obtained with graphs shown in Fig. 10.5. Strain ratio is the ratio of strains at different strain rates to reference strain obtained at strain rate 10^{-3}/s. The fractured specimens at different strain rates are shown in Fig. 10.6. The graph between the fracture strain ratio and $\ln(10^3\, \dot{\varepsilon})$ is shown in Fig. 10.7.

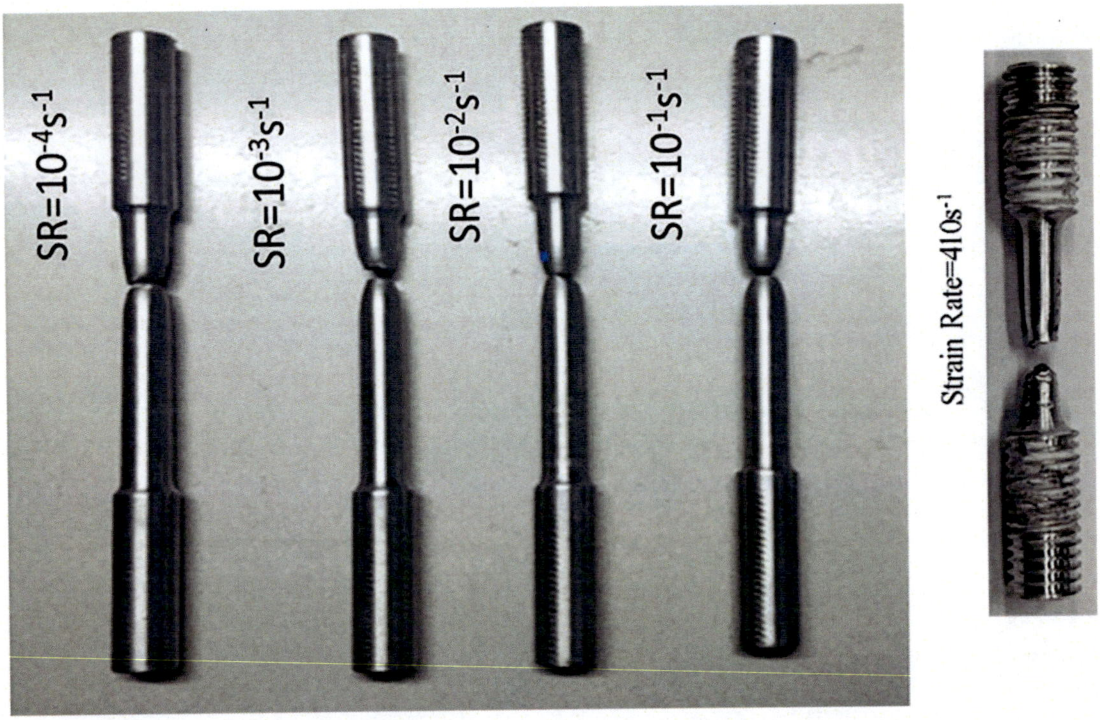

Fig. 10.6 Fractured Tensile specimens at different strain rates

Fig. 10.7 Determination of strain rate parameter (D4) for AA6063-T4

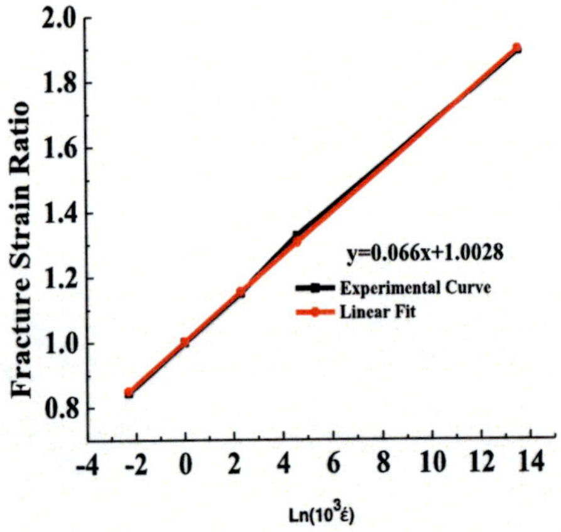

It can be seen that the fracture strain ratio is increasing with strain rates. The slope of the linear fit line obtained from Fig. 10.7 gives the value of D_4 (0.066) and is presented in Table 10.2.

The effect of temperature on fracture strain was studied by performing quasi-static tests on tensile specimens at temperatures 25 °C, 50 °C, 100 °C, 150 °C, and 200 °C at reference strain rate 10^{-3}/s. Figure 10.8 represents stress-strain curves of AA6063-T6 at different temperatures. The broken specimens after experiments at different temperatures are shown in Fig. 10.9. It can be observed from photographs of broken specimens that the ductility was increased at a higher temperature. The failure strains of specimens at different temperatures were obtained using cross-sectional areas of failed specimens. The failure strain ratio is the ratio of strain obtained at a particular temperature to strain at room temperature. The homologous temperature corresponding particular temperature is also obtained. Figure 10.10 shows the variation of fracture strain ratio with homologous temperature. A linear fit for the experimental data is constructed, and the slope of this curve gives the value of the thermal sensitivity parameter (D_5), which is 1.52. The fracture strain of the material increases with an increase in temperature due to the thermal softening of the material at higher temperatures. It is seen that the fractured diameter at higher temperatures decreases than that obtained at lower temperatures. All damage parameters D_1, D_2, D_3, D_4, and D_5 obtained with experiments are presented in Table 10.2.

Fig. 10.8 Engg stress-strain curves at various temperatures for AA6063-T6 (strain rates $10^{-3}s^{-1}$)

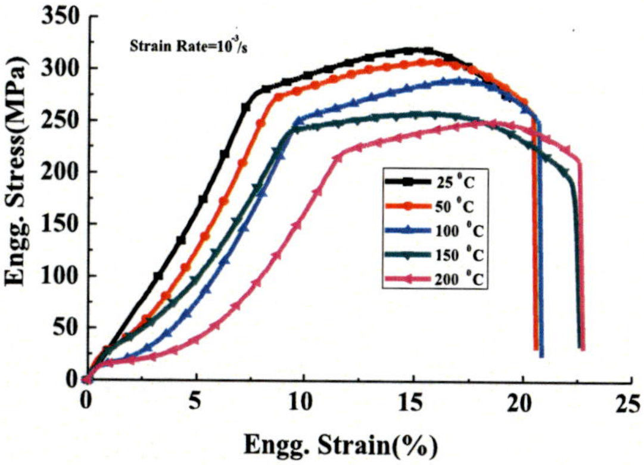

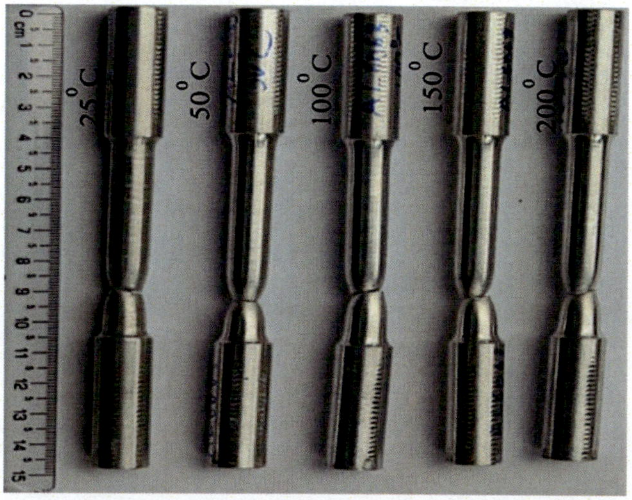

Fig. 10.9 True stress-strain curves at various temperatures for AA6063-T6 (strain rates $10^{-3}s^{-1}$)

Fig. 10.10 Determination of thermal parameter (D5) for AA6063-T6

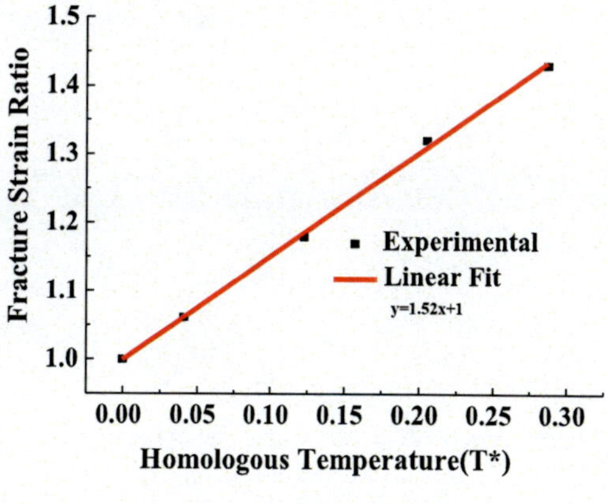

10.5 Conclusions

This paper reported the effect of strain rate and temperature on fracture strain of aluminium alloy AA6063-T6. The experimental results obtained were used to calibrate the failure model of Johnson-Cook. It was also observed that AA6063-T6 exhibited positive strain rate sensitivity and also flow stresses decreased at higher temperatures. It was observed that at lower triaxiality, the shape of fracture was cup and cone than that at higher triaxiality fractured surface. As the stress triaxiality increases, plastic deformation became restricted so that loss of ductility occurred.

References

1. KalpakJian, S., Schmid, S.R.: Manufacturing Engineering and Technology, 7th edn. Pearson Education, Melbourne (2014)
2. Smerd, R., Winkler, S., Salisbury, C., Worswick, M., Lloyd, D., Finn, M.: High strain rate tensile testing of automotive aluminum alloy sheet. Int. J. Impact Eng. **32**, 541–560 (2005)
3. Johnson, G.R., Cook, W.H: No Title. In: A Constitutive Model and Data for Metals Subjected to Large Strains, High Strain Rates, and High Temperatures. Proceedings 7th International Symposium on Ballistics, 1983, no. April, p. 1983 (1983)
4. Lesuer, D.R.: Experimental Investigations of Material Models for Ti6Al4V Titanium and 2024-T3 Aluminum. Faa, vol. DOT/FAA/AR, no. September, pp. 1–41 2000.
5. Teng, X.: High velocity impact fracture. Ph.D. Thesis, vol. 1, no. 1994, p. 330 (2005)
6. Sharma, P., Chandel, P., Bhardwaj, V., Singh, M., Mahajan, P.: Ballistic impact response of high strength aluminium alloy 2014-T652 subjected to rigid and deformable projectiles. Thin-Walled Struct. **126**(2017), 205–219 (2018)
7. Giglio, A., Giilioli, M., Manes, A.: Mechanical behaviour of Al 6061-T6 aluminium alloy under large strain and failure. In: Numerical Modeling of Materials Under Extreme Conditions, pp. 143–171. Springer, Heidelberg (2014)
8. Dabboussi, W., Nemes, J.A.Ã.: Modeling of ductile fracture using the dynamic punch test. Int. J. Mech. Sci. **47**, 1282–1299 (2005)
9. Clausen, A.H., Børvik, T., Hopperstad, O.S., Benallal, A.: Flow and fracture characteristics of aluminium alloy AA5083-H116 as function of strain rate, temperature and triaxiality. Mater. Sci. Eng. A. **364**(1–2), 260–272 (2004)
10. Brar, N.S., Joshi, V.S., Harris, B.W.: Constitutive model constants for Al7075-T651 and Al7075-T6. AIP Conf. Proc. **1195**(1), 945–948 (2009)
11. Johnson, G.R., Cook, W.H.: A Constitutive Model and Data for Metals Subjected to Large Strain, High Strain Rates and High Pressures. Seventh Int. Symp. Ballist., pp. 541–547 (1983)
12. Bridgman, P.W.: Studies in Large Plastic Flow and Fracture. McGraw-Hill, New York (1952)

Chapter 11
Measuring Adhesion Strength of an Improved Dental Biofilm Model on a Titanium Surface

M. N. Hessin, J. D. Boyd, and M. E. Grady

Abstract Adhesion of bacteria to oral implant surfaces can lead to oral infections, and the prevention of strong biofilm adherence to implant surfaces can assist in the prevention of these infections like peri-implantitis. In prior studies, single species biofilm adhesion has been quantitatively measured via the laser spallation technique. However, colonizing oral biofilms rarely consists of a single bacteria species. Multiple early colonizer species, including several strains of Streptococci, dominate initial oral biofilm formation. This study aims to characterize the adhesion of a multi-species oral biofilm consisting of *S. oralis*, *S. sanguinis*, and *S. gordonii* on titanium, a common implant material, using the laser spallation technique. Previous work has established these specific Streptococci strains as a multi-species periodontal biofilm model. This study is the first to provide a quantitative adhesion measurement of this multi-species model onto a dental implant surface. First, adhesion strength of the multi-species model is compared to adhesion strength of the single-species streptococci constituents. Fluorescent staining and imaging by fluorescent microscopy are used to identify individual bacteria species within the biofilm. The multi-species biofilm presented in this study provides a more representative model of in vivo early biofilms and provides a more accurate metric for understanding biocompatibility on implant surfaces.

Keywords Laser spallation · Adhesion · Biofilms · Dental implants

11.1 Introduction

Laser spallation technique, originally used on thin film substrates, has recently been evaluated on its effectiveness to detach cells and biological substances [4, 8]. Boyd et al. adopted the laser spallation technique to measure adhesion of single-species biofilms containing *S. mutans* [1, 2]. However, oral biofilms contain more than one single species, and initial oral biofilms on teeth are composed of mainly Streptococci species, three of which are *S. oralis*, *S. sanguinis*, and *S. gordonii* [6]. These three gram-positive streptococci species are common pioneer colonizers since they attach directly to the pellicle surface on the tooth [3, 11]. The pellicle is made up of salivary proteins that coat the teeth [6]. After early-colonizing bacteria adhere to the pellicle surface, later-colonizing bacteria attach to the early-colonizing bacteria to form a mature biofilm [7]. Strongly adhered biofilms to dental implants cause infections and can lead to implant failure. Studying the adhesion of bacteria and cells to these surfaces can potentially allow for better implant biomaterial selection that favors the adhesion of cells over the adhesion of bacteria.

Since early-colonizing bacteria form the bases of a dental biofilm, they are examined in this study as a multi-species biofilm to represent an initial dental biofilm. A stress-based adhesion measurement using laser spallation is analyzed on the multi-species model. *S. oralis*, *S. sanguinis*, and *S. gordonii* are all facultative anaerobes, signifying they can survive in both an anaerobic and aerobic environment [9]. Oral bacteria can migrate underneath the gum between the gum tissue and dental implant, which is an anaerobic environment, and can lead to implant failure. Hence, the adhesion of these bacteria cultured anaerobically is of interest. Future experimentation with an anaerobic multi-species biofilm will be done using a Sheldon Manufacturing anaerobic chamber to culture biofilms. The intention is to use the laser spallation technique on bacteria cultured anaerobically to simulate an environment close to that underneath the gums.

M. N. Hessin (✉) · J. D. Boyd · M. E. Grady
Department of Mechanical Engineering, College of Engineering at the University of Kentucky, Lexington, KY, USA
e-mail: maria.hessin@uky.edu

A. Amirkhizi et al. (eds.), *Challenges in Mechanics of Time-Dependent Materials & Mechanics of Biological Systems and Materials, Volume 2*, Conference Proceedings of the Society for Experimental Mechanics Series, https://doi.org/10.1007/978-3-031-17457-5_11

11.2 Methods and Materials

Culturing Bacteria All single species biofilms of *S. oralis* in this study were cultured planktonically for 24 h in Brain Heart Infusion (BD Difico Bacto Brain Heart Infusion; Fisher Scientific) supplemented with yeast extract (Bacto Yeast Extract; Thermo Fisher Scientific), L-cysteine (Sigma-Aldrich), hemin (Pig Hemin; Thermo Fisher Scientific), and menadione (2-mehtyl-1, 4-naphthoquinone, vitamin K3; Thermo Scientific). Per each biofilm, 1 mL of the planktonic culture was dispensed on a glass bottom petri dish at an OD_{600} of 0.7 along with 3 ml of 75 mM sucrose in BHI culture media broth. Biofilms were cultured for 24 h in an incubator at 37 C with 5% CO_2.

Staining Bacteria Each bacterial species was stained with a different nucleic acid stain to distinguish each species in the multi-species biofilm. The stains were chosen based on excitation and emission wavelengths corresponding to DAPI, FITC, and mCHERRY filter cubes. SYTO-41, SYTO-21, and SYTO-64 were chosen as the DAPI, FITC, and mCherry stains, respectively. The stains were chosen so that each stain would only be detected in its corresponding filter cube on the fluorescent microscope. Three single-species biofilms were cultured on glass petri dishes for 24 hours and then stained with either SYTO-41, SYTO-21, or SYTO-64. Each SYTO stain was mixed with PBS to a final concentration of 1 μM and dispensed over its respective biofilm. Each biofilm was then washed with 1 mL of PBS to remove excess stain. Another single-species biofilm of *S. oralis* was stained planktonically before biofilm formation. The stain was added to the planktonic culture to a final concentration of 1 μM. Once stained, the bacteria were centrifuged and resuspended in media to omit excess stain. After resuspension, the bacteria were plated on glass petri dishes and cultured for 24 h.

Imaging Images of each biofilm were taken on a fluorescent microscope at 10× magnification to validate that each stain has an emission spectrum that can only be shown in its respective filter cube. After culturing, each biofilm was aspirated to remove excess media before being analyzed under DAPI, FITC, and mCherry filter cubes on the Nikon fluorescent microscope. The brightness and exposure were kept constant when switching from one filter to the next.

Preparing Substrates for Laser Spallation Future experimentation will include biofilms cultured on titanium surfaces. The substrate will include a glass layer sandwiched between one thin layer of titanium and one thin layer of aluminum. The glass layer will provide a medium for the laser shock wave to travel, while the thin aluminum layer will absorb the energy of the laser. Water glass will be coated onto the aluminum side of the substrate to enable the propagation of the shock wave into the substrate toward the side of the titanium layer.

11.3 Preliminary Results

Three biofilms of *S. oralis* were cultured for 24 h and then stained after biofilm formation (Fig. 11.1). The first biofilm was stained with SYTO-41 blue stain and imaged with the DAPI (1a) FITC (1b) and mCHERRY (1c) filter cubes. Only in the DAPI filter cube was STYO-41 stain detected. Detection of the stain was determined based on whether the fluorescent microscope received enough signal from the fluorophore to be seen in color on the Nikon microscope software. *S. oralis* was stained with SYTO-64 red stain and imaged with the DAPI (1d) FITC (1e) and mCHERRY (1f) filter cubes. The presence of the STYO-64 stain was only detected in the mCHERRY filter cube. Another *S. oralis* biofilm was stained with SYTO-21 green stain in the DAPI (1 g) FITC (1 h) and mCHERRY (1i) filter cubes. The presence of SYTO-21 was only detected in the FITC filter cube. Due to the presence of fluorophore signal of each stain only being detected in their respective filters, Fig. 11.1 validates that the emission spectra of the stains do not overlap. After the bacteria species are stained with their respective SYTO dye and combined into a multi-species biofilm, each bacteria fluorophore will be distinguishable in their respective filter on the fluorescent microscope.

Figure 11.2 shows a biofilm of *S. oralis* stained planktonically, before biofilm formation. The key difference between Fig. 11.1 and Fig. 11.2 is the staining procedure. Figure 11.1 shows biofilms of *S. oralis* stained after biofilm formation, whereas Fig. 11.2 shows biofilms of *S. oralis* stained planktonically before the biofilm formed. Because the multi-species biofilm will include multiple stains, the bacteria species must be stained before being combined and plated. The fluorescence of the biofilm in Fig. 11.2, stained with SYTO-41 prior to biofilm formation, was only detected in the DAPI filter cube. This is congruent with the finding in Fig. 11.1 showing the fluorescence of SYTO-41 only being detected at wavelengths corresponding to the DAPI filter cube and not the FITC or mCHERRY filter cubes. Hence, each species in the multi-species biofilm will be stained with a corresponding SYTO stain, and images of each species will be visible in the fluorescent

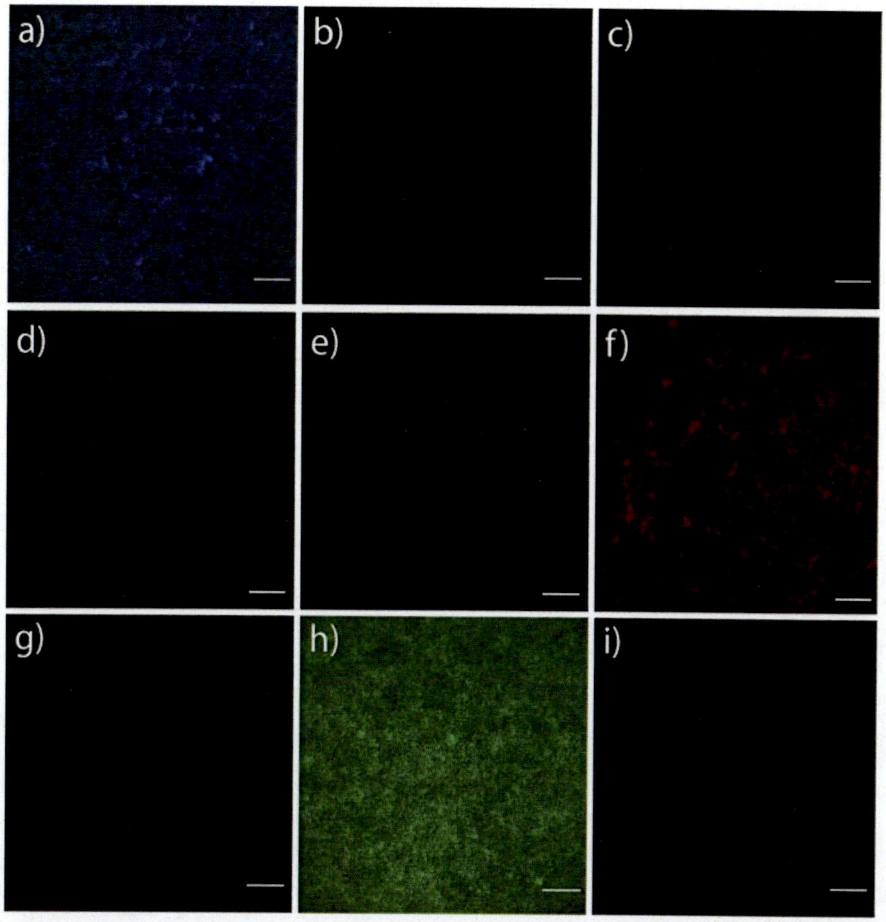

Fig. 11.1 Fluorescent microscopy images (10× magnification) of single-species aerobic biofilms of *S. oralis* stained with (**a–c**) SYTO-41 (**d–f**) SYTO-64 and (**g–i**) SYTO-21 under DAPI filter (**a**, **d**, **g**) FITC filter (**b**, **e**, **h**) and mCHERRY filter (**c**, **f**, **i**). Bacteria were stained after biofilm formation. Scale bar is 200 μm

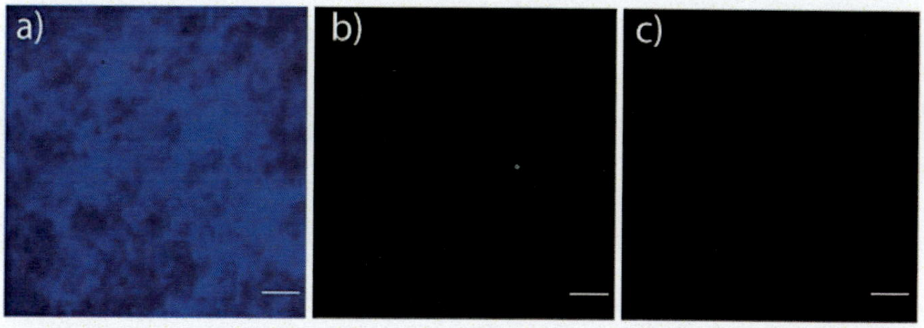

Fig. 11.2 Fluorescent microscopy images (10× magnification) of single-species aerobic biofilm of *S. oralis* under the (**a**) DAPI (**b**) FITC and (**c**) mCHERRY filter. *S. oralis* was stained with SYTO-41. Bacteria were stained before biofilm formation. Scale bar is 200 μm

microscope when the corresponding filter cube is used. Preliminary images shown in Figs. 11.1 and 11.2 are of aerobic biofilms. Future work will focus on culturing the multi-species biofilm anaerobically in an anaerobic chamber shown in Fig. 11.3a. The petri dishes will be placed in a 37 C incubator with 5% CO_2 located inside the anaerobic chamber for 24 hours to allow biofilm formation. Microscope imaging and laser spallation will be done aerobically. Future studies will use laser spallation to quantify the stressed-based adhesion of this multi-species biofilm cultured anaerobically.

The laser spallation schematic is shown in Fig. 11.3b. A high-powered, 1064 nm laser passes through a focusing lens to decrease the spot size of the laser and then is directed by a mirror optic toward the sample. The sample consists of a titanium surface with a biofilm cultured on top of the surface. The laser impinges upon the titanium-coated substrate, which sends a compressive shock wave directed through the substrate. After reflection at the free surface, the wave then loads the

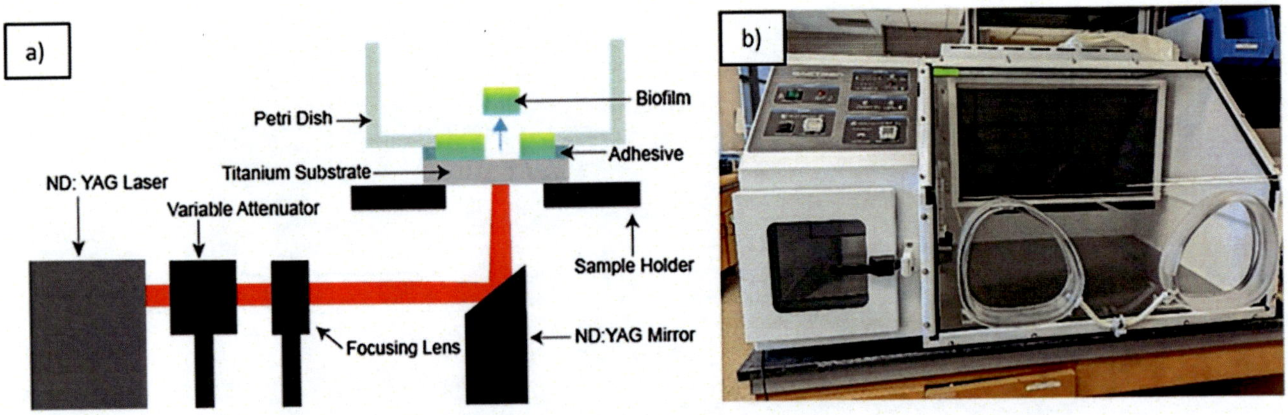

Fig. 11.3 (**a**) Laser spallation schematic and (**b**) Sheldon Manufacturing anaerobic chamber

biofilm-substrate interface in tension, which causes spalling of the biofilm from the substrate. Laser-induced shock waves are used in prior research studies to measure interfacial stress of aluminum test films [8]. Similarly, this study will capture interfacial stress between a thin film substrate and biofilm. However, due to the requirement of a vertical standing biofilm to negate effects of gravity, the interfacial stress is captured in a calibration experiment to correlate fluence with substrate stress. The fluence that is required to spall the biofilm is measured, and the stress at the titanium-biofilm interface corresponding to that fluence is recorded as the stress-based adhesion for the biofilm. A calibration experiment is performed to determine the interface stress that corresponds to each fluence. In a calibration experiment, the titanium substrate will be impinged upon by the 1064 nm laser at a known fluence. A 532 nm laser will be interferometrically set up to measure the displacement at the spot where the substrate was impinged. A photodetector will be placed in front of the fringe patterns that are created from the interferometry setup, and an oscilloscope will output voltage readings in terms of light intensity. The number of fringes can be found by unwrapping the oscilloscope voltage data [5]. The substrate stress can be determined by using the Doppler effect and 1D wave mechanics [8, 10].

11.4 Conclusion

Stress-based adhesion of a multi-species dental biofilm on a dental implant surface can be obtained through laser spallation. Adhesion of single species has been measured via laser spallation, but biofilms contain more than one single species. Streptococci are common early colonizers, and because early colonizers form the basis of initial biofilms, they are examined in this study. Preliminary data shows that SYTO-41, SYTO-21, and SYTO-64 are stains without overlapping emission spectra. Incorporating the three stains will enable the distinction of each species in the multi-species biofilm. Distinguishing constituent species in the biofilm verifies that the biofilm contains all three strains. Adhesion of a multi-species biofilm is examined in this study as a representative model of an initial dental biofilm.

Acknowledgments We gratefully acknowledge NIH Center of Biomedical Research Excellence (COBRE) in Pharmaceutical Research and Innovation (CPRI, P20GM130456) and NIH NIDCR funding (R03DE029547) for completion of these experiments.

Funding This material is based upon work supported by the National Science Foundation CAREER Award grant number 2045853. Any opinions, findings, and conclusions or recommendations expressed in this material are those of the authors and do not necessarily reflect the views of the National Science Foundation.

References

1. Boyd, J.D., Korotkova, N., Grady, M.E.: Adhesion of biofilms on titanium measured by laser-induced spallation. Exp. Mech. **59**, 1275–1284 (2019)
2. Boyd, J.D., Stromberg, A.J., Miller, C.S., Grady, M.E.: Biofilm and cell adhesion strength on dental implant surfaces via the laser spallation technique. Dent. Mater. **37**, 48–59 (2021)
3. Ebersole, J.L., et al.: The periodontal war: microbes and immunity. Periodontology. **2000**(75), 52–115 (2017)

4. Ehsani, H., Boyd, J.D., Wang, J., Grady, M.E.: Evolution of the laser-induced spallation technique in film adhesion measurement. Appl. Mechan. Rev. **73** (2021)
5. Hagerman, E., Shim, J., Gupta, V., Wu, B.: Evaluation of laser spallation as a technique for measurement of cell adhesion strength. J. Biomed. Mater. Res. A. **82A**, 852–860 (2007)
6. Howard, K.C., Gonzalez, O.A., Garneau-Tsodikova, S.: Porphyromonas Gingivalis: where Do we Stand in our Battle against this Oral Pathogen? RSC Med Chem. **12**, 666–704 (2021)
7. Huang, R., Li, M., Gregory, R.L.: Bacterial interactions in dental biofilm. Virulence. **2**, 435–444 (2011)
8. Kandula, S.S.V.: Delamination of thin film patterns using laser-induced stress waves. University of Illinois at Urbana-Champaign (2008)
9. Loo, C., et al.: Role of a nosX homolog in Streptococcus gordonii in aerobic growth and biofilm formation. J. Bacteriol. **186**, 8193–8206 (2004)
10. Wang, J., Weaver, R.L., Sottos, N.R.: A parametric study of laser induced thin film spallation. Exp. Mech. **42**, 74–83 (2002)
11. Zheng, W., et al.: Distinct biological potential of Streptococcus gordonii and Streptococcus sanguinis revealed by comparative genome analysis. Scientific Reports. **7** (2017)

Chapter 12
Time-Resolved Characterization of Impact Testing

Phillip Jannotti

Abstract Time-resolved impact studies were conducted to bridge the gap between material property testing and full-scale ballistic experiments. The motivation is to characterize the projectile-target interaction at early time scales, which includes both the target and projectile response. Prior work has focused on demonstrating the utility of in situ diagnostic techniques for investigating the physics of failure of high-strength metallic and ceramic targets; however, less work has been done to experimentally quantify the projectile deformation. Digital image correlation has been identified as a promising technique to provide real-time, full-field characterization of the dynamic failure behavior during ballistic impact. In instances where the material response is multiaxial, DIC provides one of the only practical means of collecting full-field 3D data. Stereo DIC involves applying a speckle pattern to the target surface and tracking the deformation using two synchronized cameras. This allowed for the 3D reconstruction of deformation histories. The focus of the current study is to understand the limitations and uncertainty of ultra-high-speed DIC for measuring high-rate deformation. Copper long rods were launched into ceramic anvils at a range of velocities (250-500 m/s) to characterize the axial strain and strain rate. A methodology is presented for implementing ultra-high-speed DIC in addition to concurrent multiflash X-ray imaging.

Keywords DIC · Ballistic impact · Ultra-high-speed imaging · Flash X-ray · Boron carbide

12.1 Introduction

Critical knowledge gaps remain unanswered surrounding the early-time projectile target interactions and onset of failure during ballistic impact. This is due in part to the complex nature of a ballistic event which involves dynamic loading conditions, extreme pressures, and evolving (time-dependent) multiaxial failure modes. It is also largely a consequence of inadequate in situ techniques which possess sufficient temporal and spatial resolution to characterize the high-rate deformation.

The motivation of this work is understanding how materials deform and fail during high-rate loading. This is important for guiding material processing efforts, informing full-scale testing, and supporting development of advanced computational models. Thus, tools and experiments are needed to not only visualize deformation under relevant loading conditions but also quantify the pertinent mechanics during impact and penetration. Few test techniques can adequately simulate the loading conditions experienced during impact to probe the resulting material response. Taylor impact has been explored as a suitable characterization technique because it subjects materials to very large strains and strain rates and is amenable to full-field techniques. However, Taylor tests traditionally involve a rigid target. In reality, the target and its resistance to penetration introduce complexities to the problem that can only be understood through terminal ballistic experiments. Thus, the current work employs a similar methodology to instrumented Taylor testing [1–3] with the replacement of the rigid anvil target with a lightweight ceramic anvil of finite thickness.

Over the years, a range of in situ diagnostics have been applied to Taylor impact testing including high-speed cameras, velocimetry, and digital image correlation (DIC) [1–7]. However, ultra-high-speed DIC has been identified as a valuable tool for understanding impact and penetration problems due to its ability to make full-field measurements of dynamic deformation including displacement, velocity, strain, and strain rate [8]. The basic premise of stereo (3D) DIC is to apply a semi-random speckle pattern to a sample and then track pattern deformation with multiple synchronized cameras. Recent

P. Jannotti (✉)
U.S. Army DEVCOM Army Research Laboratory, Aberdeen Proving Ground, Aberdeen, MD, USA
e-mail: phillip.a.jannotti.civ@army.mil

A. Amirkhizi et al. (eds.), *Challenges in Mechanics of Time-Dependent Materials & Mechanics of Biological Systems and Materials, Volume 2*, Conference Proceedings of the Society for Experimental Mechanics Series, https://doi.org/10.1007/978-3-031-17457-5_12

work has demonstrated the feasibility of applying ultra-high-speed DIC at 1 MHz to rod impacts [1, 2]. This investigation will seek to explore the practical limitations of DIC at 5 MHz applied to similar rod impacts on brittle targets that are subject to failure and penetration.

A methodology is presented for small-scale impact experiments of pure oxygen-free high thermal conductivity (OFHC) copper rods impacting commercial grade boron carbide (B_4C) at striking velocities ranging from approximately 250 to 500 m/s. The testing draws parallels to prior work applying DIC to Taylor impact testing but adds the complexities of ceramic failure and will utilize flash X-rays to provide insight into the ceramic penetration response. Representative experiments on B_4C will demonstrate the ability of the methodology to reveal the deformation of metallic impactors when the target anvil is allowed to fail and penetration of the rod occurs behavior.

12.2 Experimental

A highly integrated experimentation platform, high-voltage in situ diagnostic radiography apparatus (HIDRA), was utilized (see Fig. 12.1) to bridge the gap between mechanical property characterization and full-scale ballistic testing and to elucidate high-rate material behavior at early time scales. The apparatus incorporated a range of in situ diagnostics including multi-flash X-ray imaging, ultra-high-speed imaging, high-speed 3D imaging, and photonic Doppler velocimetry (PDV), although PDV is not discussed here for sake of brevity. The focus of the current study was the integration of flash X-ray imaging and ultra-high-speed DIC to investigate the early-time deformation of a long rod impactor on a lightweight ceramic anvil/target. Eight 150 keV X-ray sources were aligned radially around the target and triggered successively during the first 20 μs after impact. Two ultra-high-speed cameras (Shimadzu HPV-X2) captured 3D deformation of the metallic rod during impact and penetration at 5 Mhz with 110 ns exposure. The cameras were vertically offset and included a stereo angle of ~20°. The impact experiments were conducted using a 12.7 mm caliber smooth bore powder gun to launch 6.4 mm diameter × 32 mm long rods of OFHC copper at ~250–500 m/s. The rods were held in polycarbonate discarding sabots that separated over the 300 mm standoff from the muzzle to the ceramic impact face. Optical beam breaks near the muzzle of the gun were used for measuring the projectile velocity as well as for downstream triggering of equipment, including the two Xenon light strobes

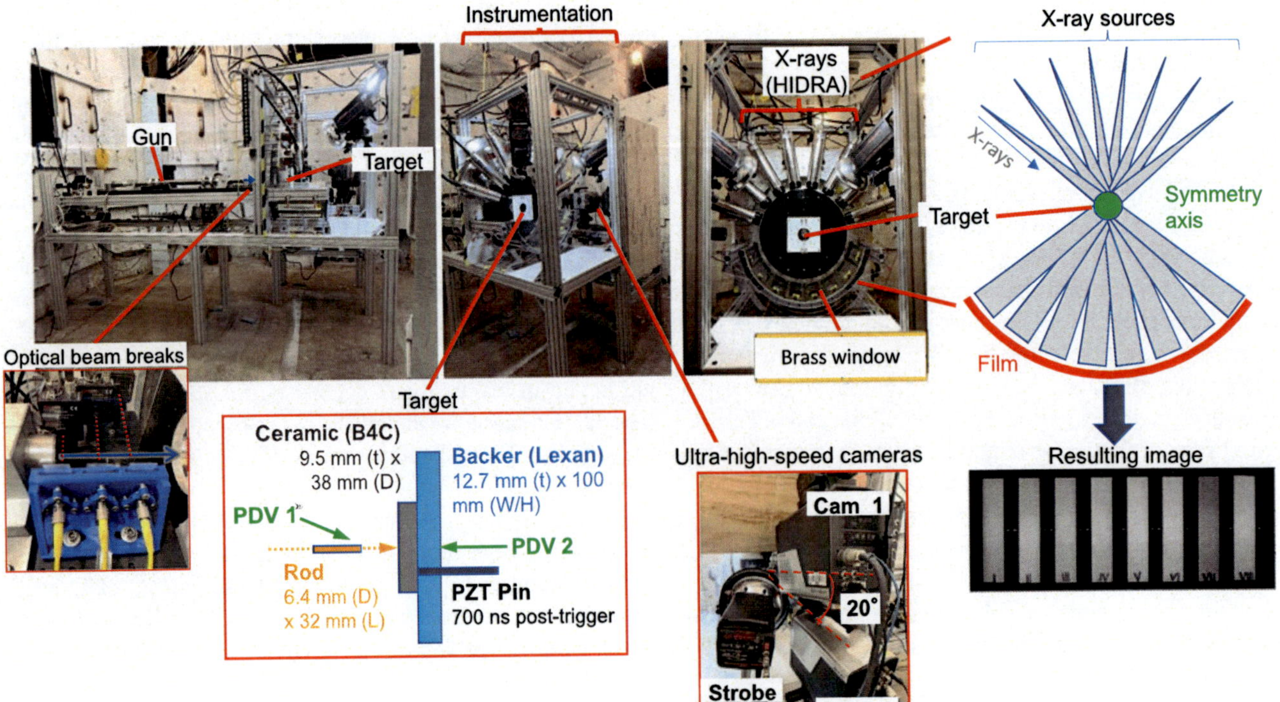

Fig. 12.1 Image of the experimental setup used for instrumented ballistic testing. Insets are included of the optical beam breaks, target configuration, ultra-high-speed optical imaging, and high-speed X-ray imaging

Table 12.1 DIC parameters for data collection and analysis

Parameter	Setting
Camera/Sensor/Digitization	Shimadzu HPV-X2 / 400x250 px / 10-bit
Lens/Imaging Distance	200mm macro, f/16 / 300 mm
Sampling Rate / Exposure	5 Mfps / 110 ns
Imaging Resolution	57 μm
ROI	23 x 14 mm
Subset Size	17 pixels
Step Size	5 pixels
Interpolation / Correlation	Optimized 8-tap / NSSD
Strain Filter	11
Calibration Panel	9 x 9 dots (0.89 mm spacing) – backlit glass
Noise (e, de/dt)	0.07%, $0.08\text{x}10^4$/s

(Photogenic PL2500DRC, 1000 W/s). It is important to note that the illumination was cross-polarized, meaning that both the light source and camera lenses utilized linear polarizers and the camera lens polarizer was rotated 90° relative to the polarizer on the light source. This dramatically reduced the effects of glare commonly occurring due to round and reflective surfaces. A lead zirconate titanate (PZT) pin in contact with the rear surface of the ceramic target was used for providing a trigger to the cameras and X-rays. Due to the very high camera framing rates (5 MHz) and short record times of approximately 25 μs, triggering needed to be reliable and precise. The typical target construction was a ceramic disk (38.1 mm diameter and 9.5 mm thick) bonded to 12.7 mm thick and 100 mm wide square polycarbonate backing plate. Although the current study only used flash X-ray imaging and DIC on the rod impactor, a polycarbonate backing enables a broader array of diagnostics to be employed and can yield a more comprehensive view of the ceramic failure/penetration response. Because the polycarbonate backing is optically transparent, additional cameras can be used to image the rear surface fracture behavior, and PDV can be employed to track the back surface velocity response.

The target rod was patterned by applying a base coat of matte white spray paint followed by manually applying dots with an extra fine-tip permanent marker. Given the resolution of the camera system (57 μm or ~ 18 px/mm), the ideal speckle size was ~170–290 μm (3–5 pixels [9]). A summary of the experimental setup and post-processing parameters is given in Table 12.1, as recommended [9]. The calibration grid was a 9×9 backlit glass grid with 0.4 mm dots and 0.9 mm spacing. Several image sequences (128 frames each) were collected at 60 Hz and were used to calibrate the stereo system. The image sequences included sufficient in-plane and out-of-plane translation and rotations to get a proper calibration while keeping the calibration grid in focus. The validity of the camera calibration was verified by an extended noise floor measurement, which involved translating and rotating the patterned target rod in the field-of-view (FOV) and then evaluating the resulting noise floor.

12.3 Results

Figure 12.2 shows representative deformation for a Cu rod impacting a ceramic target at 249 m/s, including 3D DIC data (Fig. 12.2a, b) and flash X-ray images (Fig. 12.2c). The full-field DIC data illustrates the axial strain and corresponding strain rate over the first 20 μs after impact. Note that soon after impact, the FOV was obscured by debris, and data could no longer be collected. The axial strain was maximum just below the impact face (41%), and the strain rate was found to be maximum at the impact face (6.9×10^4/s). The flash X-ray images demonstrate the extreme level of deformation taking place during the impact and penetration of the rod into the ceramic target. Unlike traditional Taylor rod impacts into a rigid anvil, there is ceramic ejecta emanating from the impact site, severe petalling, back extrusion and erosion of the metallic

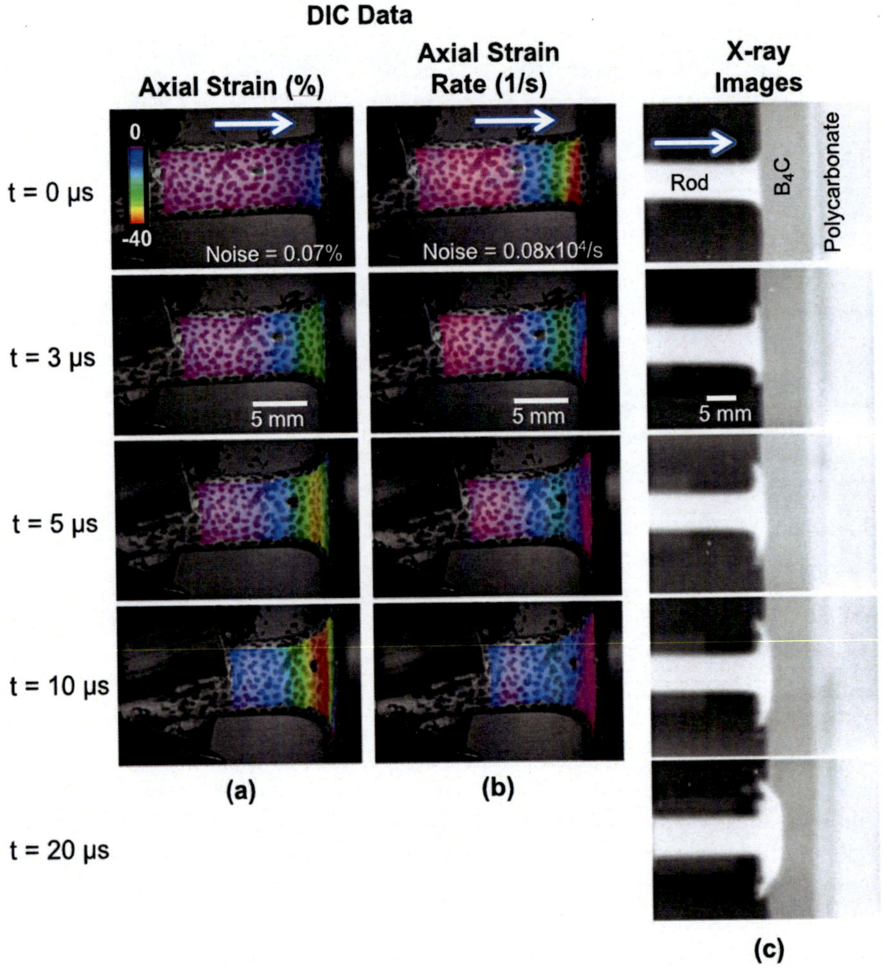

Fig. 12.2 Representative full-field DIC data showing (**a**) strain and (**b**) strain rate as well as (**c**) X-ray images of rod deformation and penetration at selected times after impact

impactor, and eventually penetration of the rod once the ceramic target resistance has been overcome. These experiments can be viewed as an extension of earlier work applying DIC to Taylor impact experiments, where in situ characterization has enabled more complex impact arrangements to be conceived since the rod deformation is being measured directly in real time. The current study was intended to gage the level of deformation and rates associated with penetrating impact events and the feasibility/limitations of this type of test. In fact, the strain and strain rate sensitivity were found to be acceptable with a strain and strain rate noise floor of 0.07% and 0.08×10^4/s, respectively. The information collected can be critical to understanding the role of loading states and rates on the deformation of metallic impactors as well as on fracture and failure of lightweight ceramic materials.

The peak axial strain and strain rates for impacts of ~250–500 m/s are summarized in Fig. 12.3. The data was compared to prior results from Taylor impact experiments using OFHC Cu rods and a rigid steel anvil [1, 2, 4]. For typical Taylor experiments at 50–200 m/s, the strain and strain rate were seen to increase rapidly from 10 to 35% and 0.1 to 9×10^4/s, respectively, with small increases in impact velocity. For ballistic impacts at 249 m/s, the target is quasi-rigid over the first 10 μs, and the rod dwells on the impact surface over the time that the strain and strain rate reach a maximum. However, the length of rod and persistent loading eventually led to ceramic failure and penetration of the metal rod (see Fig. 12.3(c)). Note that the peak strain and strain rate data fall in line with the Taylor test data (i.e., rigid anvil target). At higher impact velocities and peak pressures, the onset of penetration is observed on touchdown. Greater shock pressure and brittle fracture lead to reduced target resistance and therefore reduced strain and strain rate in the rod. Future experiments can be performed with a range of impactor materials to evaluate the role of projectile strength in the projectile-target interaction as well as the influence of stress state and fracture rate in ballistic experiments.

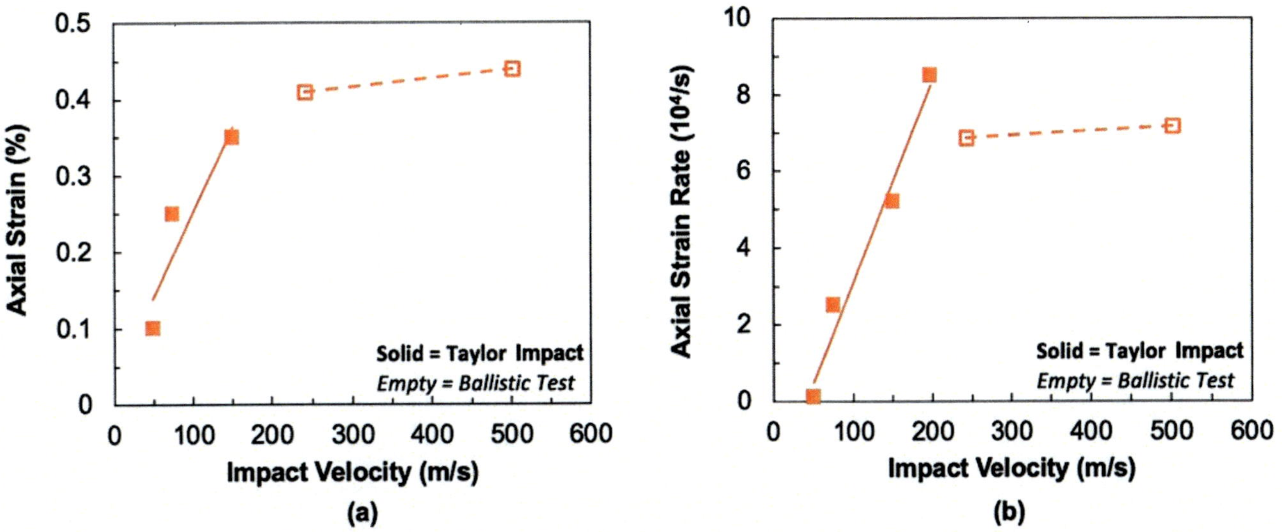

Fig. 12.3 Peak (**a**) strain and (**b**) strain rate as a function of impact velocity. Impact experiments that penetrate a ceramic target (empty data points) are compared to prior Taylor impact testing on a rigid steel anvil (solid data points)

12.4 Conclusion

A methodology was presented for conducting ballistic impact studies on OFHC copper rods coupled with concurrent ultra-high-speed DIC and high-speed flash X-ray imaging. It was demonstrated that strains of nearly 50% and 10^5/s can be captured at 5 MHz. Compared to impacts on rigid anvil target, the measured strain and strain rate of the impactor were found to be relatively insensitive to the impact velocity due to reduced ceramic target resistance at increase velocity. At sufficiently low-impact velocities, the rod experiences noticeable interface dwell and deforms/erodes at the target face for around 10 μs. As impact velocity increases, the onset of penetration occurs shortly after touchdown, and steady-state penetration ensues. Despite the high sampling rate and limited resolution, the strain and strain rate sensitivity were found to be 0.07% and 0.08×10^4/s. The instrumented ballistic testing provides promise for characterizing high-rate deformation of a metallic material (low to high strength) as long as they experience 1.5% or more axial strain such that the noise floor represents less than 5% of the measured strain. Future experiments will explore the feasibility of testing high-strength steel and tungsten carbide and further refine the approach to incorporate PDV to measure the velocity history on the rear face of the ceramic and ultra-high-speed optical imaging to monitor the fracture evolution.

Acknowledgments The author gratefully acknowledges Timothy Kerr and Eric Pierce for their assistance in performing the ballistic experiments.

References

1. Jannotti, P., Lorenzo, N., Meredith, C.: Time-resolved characterization of taylor impact testing. In: Challenges in mechanics of time dependent materials, vol. 2, pp. 63–68. Springer, Cham (2021)
2. Jannotti, P., Lorenzo, N., Meredith, C.: Application of high-speed digital image correlation to Taylor impact testing. In: Dynamic behavior of materials, vol. 1, pp. 189–195. Springer, Cham (2020)
3. Bigger, R.P., Carpenter, A., Scott, N., Dannemann, K., Chocron, S., Williams, C.: Dynamic response of aluminum 5083 during Taylor impact using digital image correlation. Exp. Mech. **58**(6), 951–961 (2018)
4. Forde, L.C., Proud, W.G., Walley, S.M.: Symmetrical Taylor impact studies of copper. Proc. R. Soc. A Math. Phys. Eng. Sci. **46**, 769–790 (2008)
5. Liu, J., Huang, F., Xu, K., Liu, L., Zuo, T., Pi, A.: Influence of mass ratio on forward and reverse ballistic impact equivalence: experiments, simulations, and mechanism analysis. Exp. Mech. **57**(3), 387–404 (2017)
6. Liu, J., Pi, A., Wu, H., Huang, F.: Dynamic response of reverse Taylor impact based on DIC technology. EPJ Web Conf. **94**, 01074 (2015)
7. Ferranti Jr., L., Gagliardi, F.J., Cunningham, B.J., Vandersall, K.S.: Dynamic Characterization of Mock Explosive Material Using Reverse Taylor Impact Experiments. In: Proceedings of the 2010 Annual Conference on Experimental and Applied Mechanic, vol. 6, pp. 337–346. Springer, New York (2011)
8. Sutton, M.A., Orteu, J.J., Schreier, H.: Image correlation for shape, motion and deformation measurements: basic concepts. Theory and Applications. Springer, New York (2009)
9. Jannotti, P., et al.: International digital image correlation society. In: Jones, E.M., Iadicola, M.A. (eds.) A good practices guide for digital image correlation (2018). https://doi.org/10.32720/idics/gpg.ed1

Chapter 13
The Effect of Aging on Delamination Strength Utilizing an Embedded Digital Image Correlation Scheme

Tomislav Kosta, Jesus O. Mares Jr., Marcel M. Hatter, Brett M. Resue, and Claron J. Ridge

Abstract We utilize an in situ/embedded digital image correlation (DIC) technique to evaluate the effect of natural aging on interface delamination strength on an idealized model system. The system consists of a single ~500–650 μm glass bead inclusion at the center of a Sylgard 184 dog-bone tensile sample along with an embedded DIC speckle pattern at the midplane of the sample. The speckle pattern enables the measurement of the evolving strain field in the region surrounding the embedded glass bead, while the sample undergoes a tensile load to failure. During the tensile loading of the sample, the applied stresses at the particle-binder interface force a debonding event to occur before the ultimate tensile failure of the sample. The measured strain field at the moment of the forced debonding event is then utilized to characterize the localized stress required to induce a failure at the interface. Through repetition of the experiment, the debonding behavior can be described in a statistically meaningful fashion. This approach is then extended to capture the effect of natural aging on the delamination behavior through a time period of approximately 24 days.

Keywords DIC · Aging · Sylgard 184 · Interfaces · Delamination

13.1 Introduction

It is well known that the bulk mechanical properties of composite materials depend highly upon the properties of the filler and matrix materials as well as the interfaces between them. Due to the high interface area in composites with a large filler volume fraction, it is important to understand the properties of the interface regions as well. We are specifically interested in the interface debonding strength in particulate composites, where the term debonding refers to the mechanical separation of two components (i.e., matrix and filler) in a composite material. Furthermore, we utilize the terms debonding and delamination interchangeably.

In the context of bulk composite strength, the interface debonding strength can play a significant role depending on the filler volume fraction. The interface debonding strength is also an important parameter in the development of high-fidelity computational models to simulate events in which material strength properties play a role. Due to the stochastic nature of failure, as well as the technical challenges involved in measuring interface debonding strength in particulate composites, it is difficult to achieve an accurate statistical representation of the debonding process.

Several standardized test methods exist for the measurement of adhesion of a simple binary material system; however, these tests are typically used as qualitative tests which provide limited quantitative data and are poorly suited for measuring the interface debonding strength in composite materials, especially those with particulate fillers. For example, tests such as the pull [1], peel [2], and blister [3] can be used to measure adhesion; however, the stress states in these tests are nonideal, and it can be difficult to interpret the effective properties associated with delamination since the data produced by these tests tends to be geometry-specific. Furthermore, these tests are not well-suited to examine particulate composites, which are the class of materials of interest in this work.

Several methods have been developed to investigate the effective interface debonding strength in particulate composites. For example, Tan et al. [4] have integrated digital image correlation (DIC) into a mode-I fracture test of a highly loaded

T. Kosta (✉) · J. O. Mares Jr. · C. J. Ridge
United States Air Force Research Laboratory, Munitions Directorate, Eglin AFB, FL, USA
e-mail: tomislav.kosta@us.af.mil; jesus.mares.2@us.af.mil; claron.ridge.1@us.af.mil

M. M. Hatter · B. M. Resue
University of Dayton Research Institute, Dayton, OH, USA

A. Amirkhizi et al. (eds.), *Challenges in Mechanics of Time-Dependent Materials & Mechanics of Biological Systems and Materials, Volume 2*, Conference Proceedings of the Society for Experimental Mechanics Series, https://doi.org/10.1007/978-3-031-17457-5_13

particulate composite to provide targeted data for cohesive zone element model parameterization. These experiments provide direct results for model calibration; however, the results are an integrated and averaged representation of the debonding process occurring at each individual filler particle. While this method provides a means of obtaining model parameter values, the statistical distribution of the properties of the individual interfaces is lost due to the integrated nature of the measurement.

On the other hand, several experiments have been developed which explicitly study the debonding of a single particle, such as those by Gent and Park [5]. Gent and Park [5] examined the debonding of a single glass sphere embedded in an optically transparent matrix material. They observed the onset of debonding during a tensile loading event and utilized analytical methods to determine the stress state in the matrix material at the debond event. They referred to this stress as the critical debonding stress and noted that it serves as an indicator of the strength of adhesion between the matrix material and glass particle.

In previous work, we present a technique [6] where we extend the methodology introduced by Gent and Park [5] to incorporate the DIC technique. Our approach utilizes an embedded DIC speckle pattern to quantify the strain field surrounding the midplane of a glass sphere embedded in an optically transparent matrix material. With this approach, we are able to directly capture the strain field during a tensile loading event up to and through a debonding event. We summarize the technique in the following section, but the reader is referred to reference [6] for a more complete discussion.

A well-known issue with polymers, and thus polymer matrix composites, is that of aging. It is well established that the properties, including mechanical properties, of polymers change as a function of time after sample fabrication. As discussed in work by White [7], aging of polymers is a highly complex process that is difficult to accurately study in totality, but age-related changes can typically be categorized as those due to physical changes (no chemical reactions), chemical changes (i.e., crosslinking), or thermal conditioning. However, it should be noted that in reality, aging is due to a combination of many mechanisms. In another work [8], we presented an aging study of Sylgard 184, where we demonstrated that the compressive elastic modulus of the material increases with time after sample production and tends toward a final asymptotic value.

Given that the properties of polymers are subject to the effect of aging, it is not unreasonable to assume that the properties of matrix-filler interfaces may change as a function of time as well. There is literature which suggests that the interface/interphase region of polymer composites is subject to the effect of environmental age in addition to the matrix and filler materials. For example, Tsotsis and Lee [9] performed various fracture tests on two fiber-reinforced composites that experienced accelerated aging via thermal conditioning and showed in both cases that as age increases, the fiber-matrix interface adhesion decreases. Hodzic et al. [10] studied the effect of water aging on several fiber-reinforced composites and concluded in their study a strong correlation between interface debonding and critical fracture toughness. In the present work, we utilized the technique presented in [6] to address the topic of interface aging in a particulate composite. We applied the technique to characterize the interface debonding strength as a function of natural aging, where "natural aging" refers to the progressive change in material properties under ambient conditions as time passes after sample fabrication.

13.2 Methods

To characterize the effect of aging on interface delamination strength, we utilized a tensile experiment with an embedded glass sphere with an in situ DIC scheme. For a complete discussion of the technique, the reader is referred to [6], but the approach is summarized here. We constructed tensile test specimens ("dog bones") of an optically transparent elastomeric matrix material with an embedded spherical glass particle at the center and a plane of optically opaque fine tracking particles at the midplane. The dog bones have a rectangular cross section of approximately 9 mm by 12 mm. We used commercially available Sylgard 184 for the matrix material, an approximately 650 μm 3 M hollow glass microsphere for the central glass bead, and 27–32 μm Cospheric black polyethylene microspheres to construct a layer of fine tracking particles which served as a speckle pattern for DIC. The optically transparent nature of the Sylgard 184 allowed for straightforward and direct observation of the debonding event, and the optically opaque tracking particles allowed for the application of DIC to quantify the strain field surrounding the glass bead up to and through the debonding event under the application of a tensile load. A diagram of a fully constructed dog bone is shown in Fig. 13.1. As detailed in [6], these dog bones are constructed in a stepwise fashion with multiple partial cure cycles in order to properly embed the glass bead and surround it with a DIC speckle pattern. Care must be exercised in order to achieve a planar field of tracking particles which are at the midplane of the glass bead; otherwise, the relevant debonding mechanisms may not be accurately captured. The sample buildup process is summarized graphically in Fig. 13.2.

Once prepared and examined for dimensional and geometric quality, each specimen was tested using a conventional load frame. A tensile load was applied at a constant speed of 0.5 mm/s in order to apply a constant engineering strain rate. To capture the evolution of the DIC speckle pattern, a camera with a resolution of 2048 by 2048 pixels was used to capture a

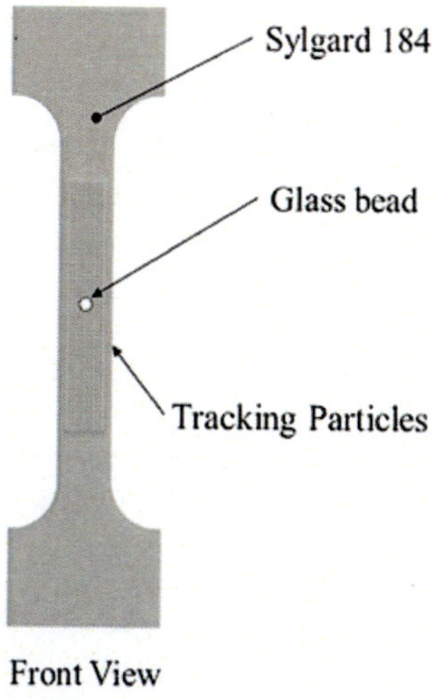

Fig. 13.1 Diagram of fully constructed tensile test sample with embedded DIC speckle pattern and glass bead inclusion [6]

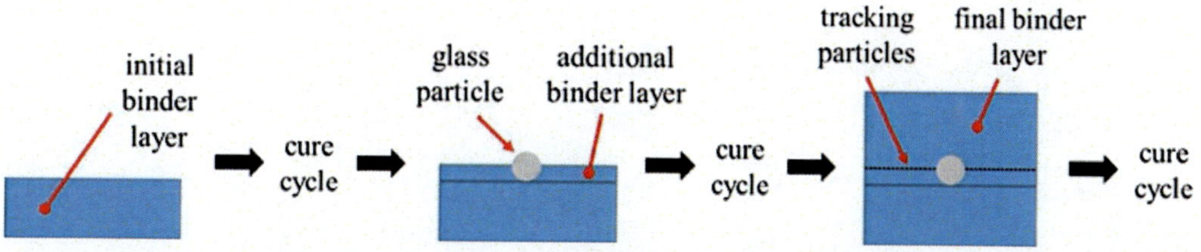

Fig. 13.2 Layered approach to constructing tensile test samples with embedded DIC speckle pattern and glass bead inclusion [6]

field of view of approximately 10 mm by 10 mm. Due to the compliance of the Sylgard 184, the camera was mounted on a translating stage to ensure the field of view of the camera includes the glass bead and a sufficient region of the surrounding speckle pattern.

For each specimen tested, the resulting data set includes a stress vs. strain curve from the load frame and a time series of images from the camera. The stress vs. strain curves provide the global stress and strain in the gage section of the dog bone. The speckle pattern images allowed for the application of DIC to obtain the time-resolved two-dimensional strain field around the glass particle inclusion. Figure 13.3 shows the strain field of the vertical component of the strain tensor, ε_{yy}, one frame pre-debond and one frame post-debond. We take the maximum local strain in the frame prior to debonding as the metric for interface strength. When combined with the mechanical properties of the matrix material, the stress state of the matrix can be calculated. The combination of the global stress condition as well as the spatially and temporally resolved local strain field provides as a rich data set describing the debonding process. This data can then be used to calibrate and validate modeling and simulation efforts to capture the process of interface debonding in particulate composites.

To characterize the effect of aging on interface adhesion, we applied the method summarized above to test the delamination behavior of sample at various time points of aging. Samples were fabricated and subsequently stored in the same controlled laboratory environment as the load frame used to conduct the tests. All samples were fabricated and tested by the same technicians, using the standardized approach described above and detailed in [6].

13.3 Results

A total of eight successful observations were measured of the debonding event for three different aging points of 73, 334, and 571 h after sample production. An example of the calculated strain field for one sample before and after the debonding event is shown in Fig. 13.3. The measured maximum local strain in the direction of tension, ε_{yy}, near the delamination event is shown for these experiments in Fig. 13.4. It is evident from the results that the maximum strain in the tensile direction required to force the delamination increases as a function of time from sample production. The average measurement increases from 0.463 mm/mm at 73 h to 0.786 at 334 h, and 0.865 mm/mm at the final time point of 571 h. This represents an increase of the average ε_{yy} debonding strain of nearly twice the value when aged from the initial measurement. It should be noted that due

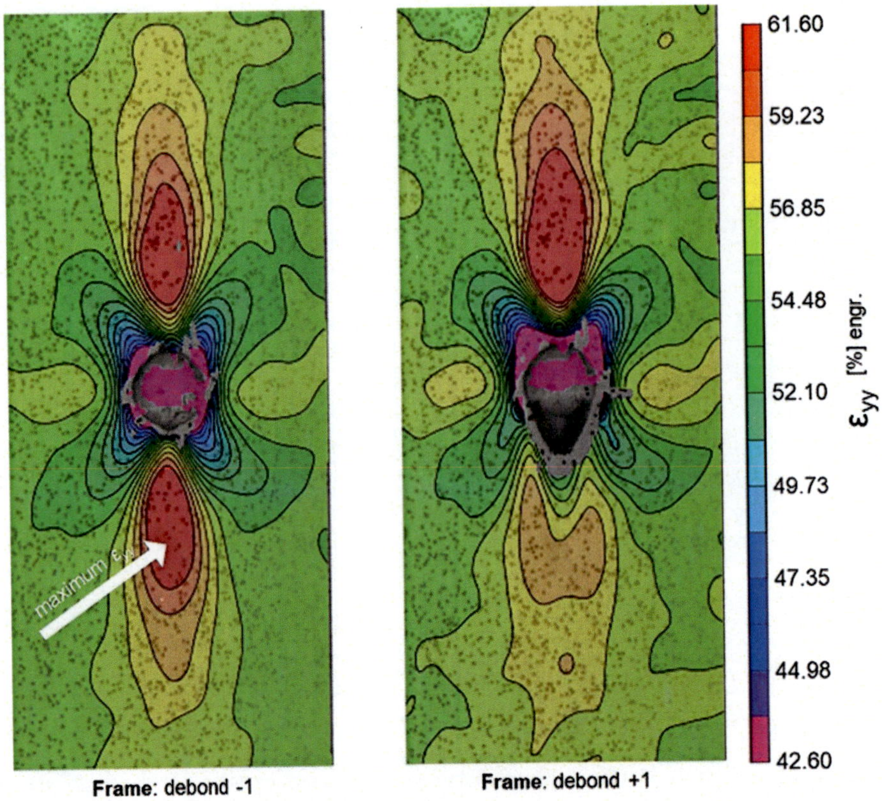

Fig. 13.3 The location of the maximum local tensile strain is indicated. Example of calculated strain field in the vicinity of the embedded particle in the vertical direction on frame pre-debond and one frame post-debond [6]

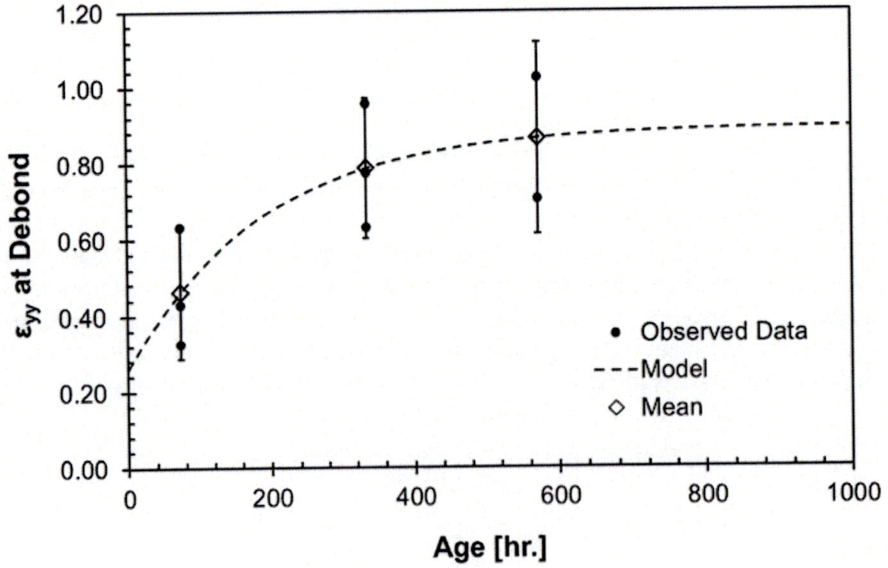

Fig. 13.4 The calculated maximum local strain in the direction of applied tension near the delamination event between the inclusion and binder. A total of eight observations were made and are displayed at each age point with the calculated mean. The displayed error bars signify 95% confidence interval of the estimated mean value. The fitted aging model is shown and indicates that the long-term debonding behavior of the sample is achieved after approximately 500 h

to the scatter of the measured data, the confidence interval of the mean value for each time point is quite large and indicates that additional observations are needed to improve the accuracy of these data.

An exponential decay growth curve was used to model the trend of the increased debonding strain as a function of time as

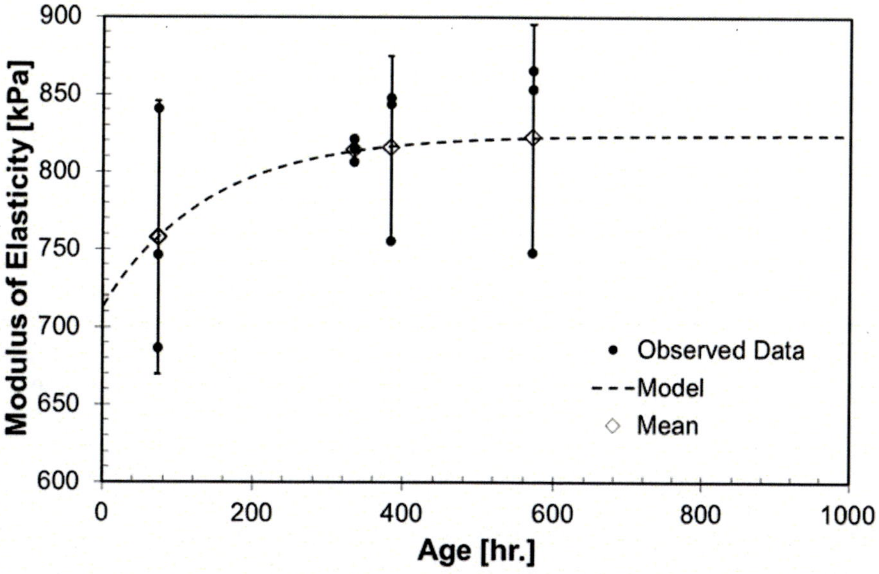

Fig. 13.5 Modulus of elasticity under tensile load of the samples as a function of age. A total of 12 observations were made and are displayed at each age point with the calculated average. The displayed error bars signify 95% confidence interval of the estimated mean value. Results show that the stiffness of the material increases with age

$$\varepsilon_t = (\varepsilon_0 + A) - Ae^{-t/t_0}$$

where ε_t is the debond strain at the time of interest, ε_0 is the initial debond strain at a time point of zero, A is a pre-exponential fitting constant in mm/mm, t is the time of interest in hours, and t_0 is a time rate fitting constant in hours. The form of the model was based upon comparative aging models [11] and was found to be in excellent agreement to capture the overall trend of the data.

The model parameters were fit using a minimization of squared error for the observed data and were calculated as ε_0 of 0.260 mm/mm, A of 0.636 mm/mm, and t_0 of 190 h. The model shows excellent agreement with the averaged results; however, additional time points should be investigated to increase confidence of the model parameters. The model indicates that the nominal debond strain is initially at a value of 0.260 mm/mm but asymptotically approaches a final debond strain of 0.896 mm/mm, representing an ultimate increase of nearly 250%. The fitted model indicates that a significant increase of the debonding strain occurs over the initial 500 h after sample production. However, at a time point of 500 h, the debond strain is within 5% of the final debond strain value. This result indicates that a minimum wait time of 500 h should be implemented after sample production before debonding experiments are to be performed.

The apparent stiffness of the material also increased as a function of age as shown in Fig. 13.5. The modulus of elasticity was determined over the bulk extension of the sample up to a strain value of 0.200 mm/mm. A total of 12 observations were measured of the modulus of elasticity under tension for 4 aging points of 73, 334, 383, and 571 h after sample production. The modulus of elasticity was measured to increase from a mean value of 758 kPa at 73 h to a mean value of 822 kPa at 571 h after sample production. The scatter of the measured data is quite large and indicates that additional observations are needed to increase the accuracy of these data. The same form of the aging model as previously described was used to fit to the observed modulus of elasticity results. The fitted parameters were calculated as K_0 of 713 kPa, A of 111 kPa, and t_0 of 141 h. The model shows excellent agreement with the averaged results; however, as with the debonding fit, additional time points should be investigated to improve the confidence of the model parameters.

The fitted model indicates that the nominal modulus of elasticity under tension is initially 713 kPa and asymptotically approaches a final value of 824 kPa, which represents an increase of approximately 15%. The results indicated that the modulus of elasticity is within 5% of the final value after a time point of 140 h after sample production. This indicates that studies on the stiffness of this material should be conducted after approximately 140 h after sample production.

13.4 Conclusions

We have applied the technique previously developed in [6] to characterize the effect of natural aging on interface adhesion in a simple binary material composite. The approach provides a unique ability to quantify the strain field surrounding an embedded particle during a tensile loading event and provides a rich data set describing the state of the matrix material leading up to and through an interface debonding event. We have utilized this technique to quantify the interface adhesion at three time points after sample fabrication. The data indicates that both the matrix properties and the adhesion behavior can both change significantly as a function of time. The most significant changes appear to occur in the initial period after sample fabrication and tend toward an asymptotic value. Additional work is needed in order to increase the confidence in the fitted model. We observed a nearly 250% increase in maximum local tensile strain at debonding over a period of approximately 500 h. This strongly suggests that age post-fabrication must be taken into consideration when conducting mechanical tests and that in order to achieve results representative of fully cured samples, one should allow material to naturally age prior to testing.

References

1. ASTM: D4541-17, Standard Test Method for Pull-Off Strength of Coatings Using Portable Adhesion Testers, ASTM Int, West Conshohocken, PA (2017)
2. ASTM D1876-08(2015)e1, Standard Test Method for Peel Resistance of Adhesives (T-Peel Test), ASTM International, West Conshohocken, PA (2015)
3. ASTM D6900-10(2015), Standard Test Method for Wet Adhesion of Latex Paints to a Gloss Alkyd Enamel Substrate, ASTM International, West Conshohocken, PA (2015)
4. Tan, H., Liu, C., Huang, Y., Geubelle, P.H.: The cohesive law for the particle/matrix interfaces in high explosives. J. Mech. Phys. Solids. **53**(8), 1892–1917 (2005)
5. Gent, A., Park, B.: Failure processes in elastomers at or near a rigid spherical inclusion. J. Mater. Sci. **19**(6), 1947–1956 (1984)
6. Kosta, T., Mares Jr., J.O.: Characterization of interface debonding behavior utilizing an embedded digital image correlation scheme. In: Lin, M.T., Furlong, C., Hwang, C.H. (eds.) Advancement of optical methods & digital image correlation in experimental mechanics. Conference Proceedings of the Society for Experimental Mechanics Series. Springer, Cham
7. White, J.R.: Polymer ageing: physics, chemistry or engineering? Time to reflect. C. R. Chim. **9**, 1396–1408 (2006)
8. Kosta, T., Krawietz, T.R., Mares Jr., J.O.: Sylgard 184 mixture and natural aging study via quasi-static compressive modulus. AIP Conf Proc. **2272**, 040007 (2020)
9. Tsotsis, T.K., Lee, S.M.: Long-term thermos-oxidative aging in composite materials: failure mechanisms. Compos. Sci. Technol. **58**, 355–368 (1998)
10. Hodzic, A., Kim, J.K., Lowe, A.E., Stachurski, Z.H.: The effects of water aging on the interphase region and interlaminar fracture toughness in polymer-glass composites. Compos. Sci. Technol. **64**, 2185–2195 (2004)
11. Kemari, Y., Mekhaldi, A., Teguar, M., Teyssedre, G.: Nonlinear regression modeling to predict thermal endurance of XLPE material under thermal aging, pp. 1–4. 2018 IEEE 2nd International Conference on Dielectrics (ICD) (2018)